Coal-Fired Generation

The Power Generation Series

Paul Breeze – Coal-Fired Generation, ISBN 13: 9780128040065
Paul Breeze – Gas-Turbine Fired Generation, ISBN 13: 9780128040058
Paul Breeze – Solar Power Generation, ISBN 13: 9780128040041
Paul Breeze – Wind Power Generation, ISBN 13: 9780128040386

Coal-Fired Generation

Paul Breeze

AMSTERDAM • BOSTON • HEIDELBERG • LONDON
NEW YORK • OXFORD • PARIS • SAN DIEGO
SAN FRANCISCO • SINGAPORE • SYDNEY • TOKYO

Academic Press is an imprint of Elsevier

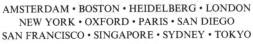

Academic Press is an imprint of Elsevier
125, London Wall, EC2Y 5AS
525 B Street, Suite 1800, San Diego, CA 92101-4495, USA
225 Wyman Street, Waltham, MA 02451, USA
The Boulevard, Langford Lane, Kidlington, Oxford OX5 1GB, UK

Notices
Knowledge and best practice in this field are constantly changing. As new research and
experience broaden our understanding, changes in research methods or professional practices,
may become necessary.

Practitioners and researchers must always rely on their own experience and knowledge in
evaluating and using any information or methods described herein. In using such information or
methods they should be mindful of their own safety and the safety of others, including parties for
whom they have a professional responsibility.

To the fullest extent of the law, neither the Publisher nor the authors, contributors, or editors,
assume any liability for any injury and/or damage to persons or property as a matter of products
liability, negligence or otherwise, or from any use or operation of any methods, products,
instructions, or ideas contained in the material herein.

ISBN: 978-0-12-804006-5

British Library Cataloguing-in-Publication Data
A catalogue record for this book is available from the British Library

Library of Congress Cataloging-in-Publication Data
A catalog record for this book is available from the Library of Congress

For Information on all Academic Press publications
visit our website at http://store.elsevier.com/

This book has been manufactured using Print On Demand technology.

Working together
to grow libraries in
developing countries

www.elsevier.com • www.bookaid.org

CONTENTS

CHAPTER *1*

An Introduction to Coal-Fired Power Generation

Coal is the most important source of electricity in the world today in terms of the amount of energy it produces. Coal plants provide around 40% of the total electricity generated globally and the proportion of electricity produced by coal-fired power plants has not changed notably over a decade. Moreover the percentage of global power produced by these plants in the second decade of the twenty-first century is marginally higher than it was during the 1970s, though not as high as it was at the middle of the twentieth century.

Coal-fired power generation is important today for a second reason. Coal combustion is one of the main anthropogenic sources of atmospheric carbon dioxide, a major contributor to global warming, and large coal-fired power plants are the most significant single point emitters of the gas. Controlling these emissions can therefore play a major role in controlling the rise in the temperature of the earth's atmosphere.

Coal has achieved its pre-eminence as an energy source for stationary power generation because it is cheap and it is plentiful. While oil remains the single most important source of global energy because of its versatility, particularly as a transportation fuel, for electricity production coal is king. The fuel is widely dispersed, though it is not found everywhere, and it is cost-effective.

Coal has been burnt to provide heat for close to 4000 years but during the eighteenth century and the Industrial Revolution its use expanded extensively for industrial applications such as smelting and for mechanical power. The modern era rise in atmospheric carbon dioxide concentrations coincides with this period and can be directly linked to the increasing combustion of fossil fuels.

Coal-Fired Generation. DOI: http://dx.doi.org/10.1016/B978-0-12-804006-5.00001-0

It was towards the end of the nineteenth century that coal began to be utilized for power generation as well as for industrial processes. Since then, most of the world's richest nations, including the USA, many in Europe, China and India have built their economic foundations on the back of coal. As a consequence this fuel is of strategic importance to many countries and this has made it extremely difficult to displace now that the dangers of global warming have been recognized. Indeed, the use of coal for power generation continues to expand and most realistic scenarios for future energy use suggest this trend is likely to continue well into the middle of the current century.

While coal is relatively cheap to mine, it does not have as high an energy density as either oil or natural gas. This makes it relatively costly to transport, so most coal is used close to the source where it is produced. There is a global market for high-quality coal but it is limited and trade routes are normally as short as possible. Most of the traded coal is used for power generation, particularly in countries such as Japan that have adopted a policy of exploiting a mixture of energy sources to increase energy security but have no coal of their own.

The traditional method of generating electricity from coal is to burn the fuel in a furnace and use the heat generated to raise steam which is then exploited to drive a steam turbine. The technologies utilized in such plants are well established and the best modern plants are capable of achieving an overall energy conversion efficiency of up to 45%. However, many older plants are less efficient than this. An alternative approach to power generation from coal is to convert the solid fuel into a combustible gas and then burn this in a gas turbine. The product of coal gasification can also be used to supply fuel to fuel cells.

Coal contains a very high percentage of carbon so the normal, uncontrolled combustion of coal in air produces large quantities of carbon dioxide which are released into the atmosphere. There are techniques available to capture the carbon dioxide from the waste gases in a coal-fired power plant. Adding carbon capture reduces the overall plant efficiency to around 30% or less, increasing the cost of electricity from such plants. Nevertheless, coal-based generation is potentially still competitive. Political and environmental pressures for change could bring carbon capture technologies to commercial maturity by the third decade of the twenty-first century. However, progress has been slow and that target looks increasingly difficult to meet.

THE HISTORY OF COAL-FIRED POWER GENERATION

The earliest established regular use of coal was around 2000 BC in northern and western China, in the region which today covers the Inner Mongolia autonomous region and the province of Shanxi. At that time wood was the fuel most commonly used for heating and cooking in most parts of the world but the availability of wood was scant in this Chinese region while coal was available close to or at the earth's surface.[1] This pattern has been copied in other parts of the world in later eras, with coal used when wood has become scarce. The use of coal in metallurgical processes and metalworking appeared much later. Greek literature cites the use of coal in metalworking at around 300 BC, the first extant reference.

While coal has been used sporadically since those times in many parts of the world, the roots of modern coal use are found in Britain, a country which has a long history of coal exploitation. Coal was mined in Britain by the Romans 2000 years ago. This coal was transported as far afield as the Rhineland where bituminous coal was traded for use in the smelting of iron ore. Coal extraction fell off after Roman influence declined during the fourth century AD and was only revived on any significant scale in the twelfth century. The fuel was used mainly for heating and cooking, replacing wood which was becoming scarce. Metalworking craftsmen, or artificers as they were then known, also took up the use of coal because it burned more hotly than wood or charcoal.

Small-scale use of coal continued into the eighteenth century when the Industrial Revolution stimulated the widespread use of coal both for industrial processes and as a fuel to drive steam engines of various types. Surface mining initially provided most coal but as the eighteenth century progressed, deep mining began to develop. Mining also developed in Germany and Poland during the eighteenth century, and to a small extent in India — then part of the British Empire — but it was not until the nineteenth century that it became widespread elsewhere. At the turn of the twentieth century Britain was still the second-largest coal producer in the world, its production exceeded only by the USA. Ironically Britain, which still has major coal reserves, is now a large importer of coal.

[1] *Use of coal in the Bronze Age in China*, John Dodson, Xiaoqiang Li, Nan Sun, Pia Atahan, Xinying Zhou, Hanbin Liu, Keliang Zhao, Songmei Hu and Zemeng Yang, The Holocene, May 2014, vol. 24, no. 5, 525–530.

The use of coal for power generation began in the USA in the 1880s, based on the same technology that was then used to create mechanical power, the steam engine. Coal was burned to raise steam and the steam used to drive an engine, which in turn drove a dynamo or alternator which produced electricity. The first fully commercial electric power station was the Pearl Street station in New York which was built by Thomas Edison and started operating in 1882. (An earlier test station was built by Edison at Holborn Viaduct, London, in early 1882 but this station only ran for a few months to prove the technology.) The Pearl Street plant used a Porter Allen reciprocating steam engine and dynamo to produce a direct current which supplied power for lighting. Output was around 130 kW.

The next major advance was the steam turbine which was invented by Charles Parsons in 1884. Steam turbines allowed more efficient energy conversion and higher outputs. By 1900 steam turbines could produce 1200 kW and by 1910 individual units could generate 30,000 kW. These large steam turbines provided the technology on which many of the power stations in the developed world relied (alongside hydropower) and during the twentieth century coal-fired power stations using steam turbines became the most important source of electricity across the globe. They remain the single most important source of electricity in the second decade of the twenty-first century.

GLOBAL ELECTRICITY PRODUCTION FROM COAL

The amount of electricity generated from coal has increased continuously since the beginning of the twentieth century. Early global figures are difficult to find but, in 1949, UN statistics indicate that 531 TWh of electric power was generated from thermal power stations; the majority of these would have been based on coal combustion along with some oil (gas was not yet widely used for power generation). Total global production of electricity in 1949, based on UN data, was 840 TWh, with thermal power plants responsible for 63% of the total, and hydropower stations for most of the remainder.[2]

By 1973 total annual global production had risen to 6117 TWh of which 2343 TWh or 38% was produced from coal and peat. These figures can be found in Table 1.1 which also shows the global production of electricity from coal between 2004 and 2012, based on data from the

[2]*World Energy Supplies in Selected Years, 1929–1950*, United Nations 1952.

Table 1.1 Annual Electricity Production from Coal and Peat[3]		
Year	Annual Production of Electricity from Coal (TWh)	Total Global Annual Electricity Production (TWh)	Coal and Peat as a Proportion of Total Annual Production (%)
1973	2343	6117	38.3
2004	6994	17,450	39.8
2005	7351	18,235	40.3
2006	7755	18,930	41.0
2007	8228	19,771	41.5
2008	8263	20,181	41.0
2009	8119	20,055	40.6
2010	8698	21,431	40.6
2011	9144	22,126	41.3
2012	9168	22,668	40.4
Source: International Energy Agency.		

International Energy Agency. Electricity production from coal in 2004 was 6994 TWh, more than the total global electricity production in 1973, and this production accounted for 39.8% of global output. Production from coal rose virtually continuously between 2004 and 2012 (the exception being 2009 during the global crisis) so that for 2012 total production of electricity from coal was 9169 TWh, 40.4% of all the electricity produced across the world.

The use of coal for power production is greatest in countries which have large reserves of the fuel. The national league tables are dominated by China and the USA. Table 1.2 contains figures for the top 10 countries listed by the volume of coal-based electricity each generated in 2012. At the top of the table is China which generated 3785 TWh from coal that year, followed by the USA with 1643 TWh and India with 801 TWh. The leading European nations were Germany which derived 287 TWh from coal and the UK with 144 TWh. It is notable that Japan and South Korea, countries that have limited coal reserves themselves, are also big producers of electricity from coal. Both must import the coal they burn.

Many of the countries in Table 1.2 have a mixture of generation types, but some rely heavily on coal for their electric power. This is illustrated in Table 1.3 which shows the 12 countries that relied the

[3]IEA Key World Energy Statistics 2006–2014.

Table 1.2 National Production of Electricity from Coal, Top 10 Leading Nations in 2012[4]

Country	Production from Coal and Peat in 2012 (TWh)
China	3785
United States of America	1643
India	801
Japan	303
Germany	287
South Korea	239
South Africa	239
Australia	171
Russian Federation	169
United Kingdom	144
Rest of the world	1387
Source: *International Energy Agency.*	

Table 1.3 Proportion of Coal in Total Electricity Production in 2012: Top 12 Nations[5]

Country	Electricity Production from Coal (%)
Mongolia	95
South Africa	93
Poland	83
China	81
India	71
Australia	69
Israel	61
Indonesia	48
Germany	44
United Kingdom	39
United States of America	38
Japan	21
Source: *World Coal Association.*	

[4]IEA Key World Energy Statistics 2014.
[5]Coal Facts 2014, World Coal Association.

most heavily on coal for their electric power in 2012. Top of the list is Mongolia which derived 95% of its electric power from coal, followed closely by South Africa with 93%. Both have extensive indigenous supplies of coal. Poland with 83% is the next nation on the list. This European country has significant lignite reserves which is uses to produce its electricity. Two of the three biggest coal energy exploiters, China (81%) and India (71%), are fourth and fifth in the table by this measure. However, the third largest exploiter, the USA, is only eleventh in the table. The USA derives 38% of its power from coal and the remainder from a mixture of other sources. Of the other countries in the list, Australia and India have extensive coal reserves. However Israel, Germany, the UK, and Japan are all importers of coal for power generation.

CHAPTER 2

Coal Types and the Production and Trade in Coal

Coal is a type of fossil fuel that contains the remains of vegetation that grew on the earth's surface during prehistoric times. This prehistoric vegetation absorbed energy from sunlight and used it to drive the photosynthesis reaction which converts carbon dioxide and water into glucose, the first stage in the cycle converting solar energy into chemical energy. In consequence, coal can be viewed as a form of stored solar energy.

Most plants decay when they die and are recycled through the soil. However under certain conditions the plant material does not decay but both builds up and become buried. It is this material, after hundreds of millions of years, that becomes coal and other fossil fuels. The type of coal that is produced depends on a variety of factors including the age of the deposit, the type of vegetation that has been buried and the temperature and pressure experienced by the vegetation at the depth at which it is buried.

Coals are ranked according to the amount of carbon they contain and their water content. High-ranking coals have a high carbon content and low water content, while low-ranking coals contain less carbon and high levels of moisture. The ranges of each of these for the main coal classes are shown in Table 2.1. The proportion of volatile material also varies from coal to coal. The physical properties of the different coals affect their suitability for use in power plants.

The hardest of coals is anthracite. This coal contains the highest percentage of carbon (up to 92% dry content) and relatively little volatile matter or moisture. When burnt it produces little ash and relatively low levels of pollution (excluding carbon dioxide). Its energy density is generally higher than other coals at 32–33 MJ/kg. Anthracite is typically slow-burning and often difficult to fire in a power station boiler unless it is mixed with another fuel, and it has traditionally been used

Coal-Fired Generation. DOI: http://dx.doi.org/10.1016/B978-0-12-804006-5.00008-3

Table 2.1 Coal Types[1]				
Coal	Dry, Carbon Content (%)	Moisture Content Before Drying (%)	Dry, Volatile Content (%)	Heat Content (MJ/kg)
Anthracite	86–92	7–10	3–14	32–33
Bituminous coal	76–86	8–18	14–46	23–33
Sub-bituminous coal	70–76	18–38	42–53	18–23
Lignite	65–70	35–55	53–63	17–18
Peat	<60	75	63–69	15
Source: Coal Marketing International.				

for heating rather than industrial use. However, it is becoming more common as a power plant fuel in countries with large reserves, such as Russia and Ukraine, which are switching to anthracite for power generation to free other fuels such as natural gas for export.

While anthracite reserves are important, the most abundant of the coals are the bituminous coals. These coals contain significant amounts of volatile matter. When they are heated they form a sticky mass, from which their name is derived. Bituminous coals normally contain between 76% and 86% carbon. Moisture content is between 8% and 18% and they may contain up to 46% volatile matter by dry weight. They burn easily, especially when ground or pulverized. This makes them ideal fuels for power stations. Bituminous coals are further characterized, depending on the amount of volatile matter they contain, as high (31–46%), medium (22–31%), or low volatile (14–22%) bituminous coals. Some bituminous coals contain high levels of sulfur which can be a handicap for power generation purposes.

A third category, called sub-bituminous coals or soft coals, are black or black-brown. These coals contain between 70% and 86% carbon and 18% to 38% water, even though they appear dry. Volatile content is 42–53%. They burn well, making them suitable as power plant fuels and the sulfur content is low.

The last group of coals that is widely used in power stations is lignites. These are brown rather than black and have a dry carbon content of 65–70%, although this is much lower as mined due to the moisture content of 35–50%. Lignites are formed from plants which

[1]http://www.coalmarketinginfo.com/advanced-coal-science/

were rich in resins and contain a significant amount of volatile material, typically 53–63% when dry. The amount of water in as-mined lignite, and its consequent low carbon content, makes the fuel uneconomic to transport over any great distance. Lignite-fired power stations are usually found adjacent to the source of fuel which needs drying before being burnt in the plant furnaces.

A type of unconsolidated lignite, usually found close to the surface of the earth where it can be strip-mined, is sometimes called brown coal. (This name is common in Germany.) Brown coal has a moisture content of around 45%. Peat is also burned in power plants, though rarely. Peat has a carbon content of under 60% and a moisture content of around 75%. Volatile material makes up 63–69% of the dry weight.

COAL RESOURCES AND THE COAL TRADE

Coal is one of the most widely available fuels. It is found in more than 70 countries according to the World Energy Council in its *2013 Survey of World Energy Resources* and is actively mined in 50 countries. This means that most nations either have access to domestic supplies or can purchase it from a close neighbor. Even so, there is an active global trade in coal and this has expanded significantly over the last 20 years.

The amount of coal available for human exploitation is usually quantified in terms of the proven reserves from different countries and regions. This figure represents the known accessible reserves. The total quantities within the earth may be significantly higher. At the end of 2013, the total proven reserves were 891,531 Mtonnes as shown in Table 2.2.

Table 2.2 also shows a breakdown of proven reserves by region. The largest reserves, 310,538 Mtonnes, are found in Europe and Eurasia. However, 70% of this coal is sub-bituminous and lignite and only 30% is harder coals such as anthracite and bituminous coal. The second-largest regional reserves, 288,328 Mtonnes are found in the Asia Pacific region, which includes China and India, both of which have large deposits. The coal in this region is roughly 55% anthracite and bituminous coal and 45% sub-bituminous and lignite. North America has total reserves of 245,088 Mtonnes, with 54% lignite and sub-bituminous and 46% harder coals. The two other regions in Table 2.2, Central and South America, and the Middle East and Africa, both have much smaller reserves.

Table 2.2 Proven Coal Reserves at the End of 2013[2]				
Region	Anthracite and Bituminous Coal Reserves (million tonnes)	Sub-bituminous and Lignite Reserves (million tonnes)	Total Reserves (million tonnes)	Reserve/Production Ratio
North America	112,835	132,253	245,088	250
Central and South America	7282	7359	14,641	149
Europe and Eurasia	92,557	217,981	310,538	254
Middle East and Africa	32,722	214	32,936	126
Asia Pacific	157,803	130,525	288,328	54
World total	404,199	488,332	891,531	113
Source: BP.				

The current level of coal global reserves is predicted to last for 113 years at the rate of extraction found across the world today. However, there are large regional variations. European and North American reserves are expected to last for around 250 years, while those in the Asia Pacific region are only expected to last for 54 years, a reflection of the vastly differing rates of consumption.

While coal reserves are widely scattered, five countries hold between them 75% of all the deposits. These are the USA (28%), Russia (18%), China (13%), Australia (9%), and India (7%).

Part of the reason for the rapid decline in Asian Pacific coal reserves can be seen in Table 2.3 which lists the top coal-producing nations of the world in 2013. At the top of the list, with by a wide margin the largest global production, is China with 3561 Mtonnes or close to 46% of the global total. The second-largest producer, the USA, mined only 904 Mtonnes or roughly one quarter of that mined by China. India, with production of 613 Mtonnes, has the third-largest industry.

Much of the production listed in Table 2.3 is for steam coal but several countries also produce significant quantities of lignite which is used exclusively in local power station. The top lignite producers are Germany, Russia, the USA, Poland, Turkey, and Australia.

[2]BP Statistical Review of World Energy June 2014.

Table 2.3 Coal Production by Country in 2013[3]	
Country	Coal Production in 2013 (million tonnes)
China	3561
United States of America	904
India	613
Indonesia	489
Australia	459
Russian Federation	347
South Africa	256
Germany	191
Poland	143
Kazakhstan	120
Rest of the world	740
World total	7823
Source: International Energy Agency.	

Table 2.4 Energy Densities for Different Fossil Fuels	
Fossil Fuel	Average Energy Density
Coal	24 MJ/kg
Crude oil	46 MJ/kg
Natural gas	54 MJ/kg

The global trade in coal accounts for about 15% of the total coal consumed according to the World Energy Council, with most coal being used in the country where it is produced. Coal has a lower energy density that either oil or natural gas, as shown in Table 2.4. The average energy density of coal is around 24 MJ/kg (although as seen above some coals can have an energy density of up to 33 MJ/kg). In contrast crude oil has an average energy density of 46 MJ/kg and natural gas 54 MJ/kg. This means that coal is more costly to transport than either of the other two fossil fuels. Transportation costs therefore account for a large part of the cost of coal so that according to BP, in 2013, imported Japanese steam coal was 56% more expensive than US Central Appalachian coal.

[3]IEA Key World Energy Statistics 2014.

As a consequence of the cost, the steam coal market (coal for power generation) is divided into two regional markets. One, the Atlantic market, comprises importing nations of Western Europe, principally the UK, Germany, and Spain. The second, the Pacific market, serves the major Asian importers, Japan, South Korea, Taiwan, India, and China.

The largest exporters of coal are Australia and Indonesia. Both are well placed to serve this Pacific market. Russia is another large coal exporter. Its traditional market has been Europe but the country is trying to win a greater share of the Pacific market. The USA, Colombia, and Canada all export coal, much of it to Europe. South Africa is also a major exporter. It is well placed to export to both the Pacific and the Atlantic markets.

COAL PROCESSING AND CLEANING

Processing coal before combustion offers a way of improving the quality of a coal, both economically and environmentally. Processing can be carried out physically or chemically but the only type of cleaning exploited commercially is physical cleaning. Chemical cleaning processes are under development but have not so far proved cost-effective.

The most well-established methods of coal-cleaning focus on removing excess moisture from the coal, reducing the amount of incombustible material which will remain as ash after combustion and reducing the sulfur content of the coal. Moisture removal reduces the weight and volume of the coal, rendering it more economical to transport and easier to burn. Ash removal also reduces the mass and volume of coal and improves its combustion properties and aids power plant performance.

Moisture is removed from coal by drying. This can simply be solar drying, leaving the coal in the open before transporting it. Drying coal in this way can reduce its mass and increase its energy density, making it relatively cheaper to transport.

The alternative, drying coal by heating, is most often carried out at the power station utilizing waste heat in the plant flue gases. Such a procedure is absolutely essential when burning high-moisture lignites such as brown coal. In this case drying does not affect transportation costs because the fuel has, by this stage, already reached the power station.

Physical separation of ash from coal relies on a difference in density between coal and its common impurities. This is applied quite widely during coal preparation, both before and at a power station.

Ash removal is carried out by crushing the coal into small particles. Incombustible mineral particles are more dense than the coal and can be separated using a gravity-based method, often with air or water as the separation medium. Fluid-based separation may exploit up-currents of air that pass through the crushed coal, sometimes in a fluid-ized bed. The lighter coal particles are carried away by the fluid for recovery at a later stage, leaving the impurities behind in the fluidized bed. Mixing the crushed coal with water also offers an effective means of separation as the coal particles float and the heavier mineral parti-cles sink. Such treatment will remove some minerals containing sulfur, and can result in a reduction of up to 40% in sulfur dioxide emissions during combustion. (Some sulfur is bound to the carbon in the coal. Such sulfur is not affected by this type of cleaning.) Coal that is cleaned using a wet cleaning process (coal washing) must then be dried before combustion.

The severity of the physical processing is associated with the fine-ness of the coal. Coal crushing or grinding is needed to release the impurity particles from the fuel. Treatment of fine coal particles tends to be more intense than that of coarse particles. Both processes involve some coal losses, which can vary from between 2% and 15% of the total coal. Rejected coal can still be used in a plant capable of burning low-quality coal such as a fluidized bed furnace.

In addition to the common physical techniques outlined above there are some more specialized techniques being developed. These including using liquid carbon dioxide to selectively trap coal particles from a slurry of coal particles in water, and an electrostatic process in which the differing electrical insulation properties of coal and mineral impuri-ties allow them to be separated after a charge has been applied to both. Cyclones that can selectively separate coal from pyrites, one of the main sources of sulfur in coal, have also been tested, as have mag-netic techniques for separation.

There have been attempts to develop more advanced methods for coal treatment employing either higher-temperature processing of the coal or chemical rather than physical processes. These processes, which are aimed at removing polluting impurities in the coal to make it a cleaner fuel to burn, have not so far found commercial application. Coal cleaning processes are also being developed to treat coal wastes that have previously been discarded to make them suitable for combustion.

Chemical cleaning processes often target specific impurities, particularly sulfur. Selective oxidation of the sulfur using specific reagents has been tested, as has a process in which the ground coal is fed into a bath of molten sodium and potassium hydroxide. The latter dissolves much of the impurity material, leaving the coal which can be separated physically from the mixture. Chemical processes such as these and others are feasible but complex and currently appear to be too costly to implement. However they may become cost-effective if they can clean the coal to a sufficient level of purity that no further cleanup to remove sulfur — for example — is required.

Dry coal cleaning was popular in the USA from the 1930s until 1990 but declined after that date. However, there has been a recent revival in interest in coal cleaning as a way of reducing transportation costs. There is also interest in India where some coals are transported up to 1000 km or more. Dry cleaning processes can reduce the ash content in some of these coals from 40% to 30%.

CHAPTER 3

Coal-Burning Technology

The first commercial power station, Thomas Edison's Pearl Street station in New York, which opened in 1882, contained all the essential elements that go to make a modern coal-fired power station. The power station burned coal in a furnace, the heat from which was exploited to produce steam in a boiler. This steam was then used to drive a steam engine, in this case a 130 kW reciprocating Porter Allen steam engine, one of the most efficient of its day. The engine changed the heat energy in the steam into a rotary mechanical motion and used this to drive one of Edison's own "jumbo" dynamos which produced a direct current output of up to 100 kW, sufficient to power 1200 of the inventor's own incandescent light bulbs. In fact, the station was equipped with six engines and six dynamos and operated virtually continuously until a fire damaged it in 1890.

The reciprocating steam engine had virtually reached the pinnacle of its development in the early 1880s and it was soon overshadowed by the steam turbine which was invented by Charles Parsons in 1884. His first steam turbine drove a 7.5 kW dynamo but size quickly increased and this technology soon replaced the reciprocating engine as the main engine, or prime mover, in a coal-fired power station. From that point on the primary elements – furnace and boiler, steam turbine, and dynamo or alternator[1] – have remained virtually unchanged, functionally, although all have become more sophisticated since the end of the nineteenth century. There have been additions since that time too. The most notable are emission control facilities that clean the exhaust

---

[1]At the end of the nineteenth century there was competition between direct current supplies, such as that provided by Thomas Edison's dynamo, and alternating current supplies, which were produced by an alternator such as that designed by George Westinghouse. Eventually, the alternating current system became the standard for transmission and distribution of electric power because it was cheaper to implement over long distances. This was because the voltage of an alternating current supply could easily be increased or decreased using a transformer, allowing high-voltage transmission over long distances and then local low-voltage distribution to users. This was not possible with direct current supplies. Today the unit that converts rotary motion into electrical energy is usually called a generator.

Coal-Fired Generation. DOI: http://dx.doi.org/10.1016/B978-0-12-804006-5.00009-5

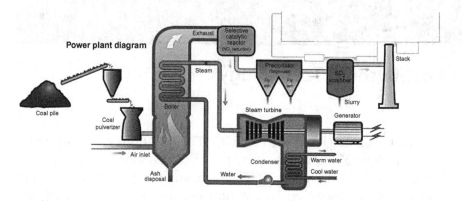

Figure 3.1 A simplified schematic of a pulverized coal-fired power plant. Source: Image courtesy of Asco.

gases exiting the furnace and boiler before they are released to the atmosphere. Fuel handling has also changed dramatically from the days when it relied on men with shovels to feed the fuel into the furnace.

The most common type of coal-fired power station in operation today is called a pulverized coal-fired plant because it burns coal that has been ground into a fine power. A simplified schematic of this type of plant is shown in Figure 3.1. Around 90% of all coal-fired power stations operating today are based on this design. The key elements of such a power plant are: the coal-handling facility which receives and stores the fuel and then prepares it for combustion; the boiler, which today is a single unit combining the furnace and the systems for converting the heat generated by the combustion into steam; the steam turbine which turns the heat energy within the steam into mechanical energy; the generator which converts the mechanical energy into electrical energy; the flue gas treatment systems which clean the waste gases before they are released into the atmosphere.

Coal will usually be delivered to a power plant as mined, unless some mechanical cleaning has already taken place. At the plant some physical sorting will take place before the coal is crushed and milled ready to be injected into the furnace. Inside the furnace, the coal enters a combustion chamber where it burns in the presence of controlled quantities of air to ensure low pollutant production. The heat energy released by the combustion process is then captured in arrays of water tubes that are placed in the path of the hot combustion gases and around the furnace walls so that all radiant and convective heat is collected. Depending on the boiler design, ash may all be carried away with the flue gases from the plant or some may collect at the bottom of the boiler enclosure where it can be extracted and treated for disposal.

Steam exits the boiler at a high temperature and a high pressure. The exact level of each will be determined by the design of the boiler. This steam is then fed into the plant steam turbine. Pulverized coal power plants are normally very large, with generating capacities of up to 1000 MW, sometimes more, and the steam turbines for these plants are made up of multiple turbine units and complex steam flows designed to extract the maximum amount of energy. The turbine units are normally divided into three groups, high-pressure steam turbines, intermediate-pressure steam turbines, and low-pressure steam turbines. The steam exiting the low-pressure steam turbines is finally condensed and returned to the boiler and recycled through the system. There may be a single generator driven by all the turbines or the multiple turbines may drive more than one generator.

The flue gases that exit the boiler after combustion and heat recovery are rich in carbon dioxide and laden with other impurities. These impurities include sulfur dioxide, nitrogen oxides, heavy metals, organic compounds, and tiny ash particles. All these impurities must be removed before the flue gas can be released into the atmosphere. Cleaning is carried out in a sequence of flue gas-cleaning systems, each a separate chemical or filtration plant. Carbon dioxide in not currently removed from commercial coal-fired power plants but technology for its capture is at an advanced stage of development and is expected to be deployed within a decade.

A plant that applies all the available emission control processes, including carbon dioxide capture, may be called a zero emission plant although in fact traces of all the pollutants will still be released. However its environmental impact will be much lower than that of an uncontrolled plant.

Efficiency is the key to modern coal-burning technology and to its future. The higher the ratio of electrical energy produced by the plant to the chemical energy input (the energy content of the coal), the cheaper each unit of electricity produced will be. For modern plants without carbon dioxide capture, higher efficiency also means lower emissions per unit of electricity produced. Alternatively, if carbon capture technologies are added to a plant, with a commensurate loss in efficiency, a high-efficiency plant will potentially still generate electricity economically.

When attempting to maximize the efficiency of a coal-fired power station, some elements of the energy conversion process can be altered by design. Others cannot. Of all the chemical energy contained with the coal, around 15% is lost during the primary energy conversion process, combustion. Much of this is irretrievable. The remainder is released as heat energy so that the hot steam produced by the boiler contains around 85% of the original chemical energy. Converting the hot steam into electricity relies on the Carnot thermodynamic cycle. Conversion efficiency depends on the temperature and pressure of the steam (more accurately the temperature and pressure drop that is achieved between steam turbine inlet and outlet) so the development of modern coal-fired power plant technology is directed at producing steam at the highest temperature and pressure possible. From an energy viewpoint, therefore, the two most important components of a coal-fired power station are the boiler which produces high-temperature, high-pressure steam and the steam turbine which must then convert the energy carried by that steam into electrical energy. However, each of the components in a pulverized coal plant plays a vital role, from the coal-handing systems to the waste gas cleanup system.

COAL HANDLING

Coal destined for combustion in a pulverized coal-fired power plant is delivered to the power station directly from the mine. If the supplier is domestic, then the delivery will normally be via road, rail, or sometimes by barge if suitable rivers or canals are available. Coal purchased on the global market will be shipped to the country of destination in bulk carriers.

However delivered, the coal will normally be off-loaded and moved by conveyor to the stockyard (or coal pile) where the local supply is stored. The chunk size and condition of the coal arriving at the plant will be determined by the method of mining and transportation.

While some plants will take coal from a single source on a long-term contract, modern market conditions mean that many plants today will receive coal from a variety of different sources. These coals will vary in heat content and the level of impurities, including ash residues, sulfur, and some trace metals, that they contain. Modern high-performance power plants are optimized for coals of a particular specification so operators must keep track of coal consignments in the stockyard and where necessary either mix coals of different quality or alter plant operating conditions to maintain optimum performance. Modern monitoring and simulation software can keep track of the quantity and quality of different coals stored in the plant stockyard and provide visualization to aid management. It should also be able to monitor the moisture content of the stored coal which will vary with levels of rainfall since the stockyard is normally exposed to local weather conditions.

From the stockyard, coal is moved through a system of hoppers and conveyors into the coal processing system. Depending on the size of the power plant there may be several independent processing lines, each feeding one or more of the furnace burners. Coal is first broken down in a coal breaker into pieces of around 2 cm in diameter, though the size will vary depending on the specific plant machinery. The broken coal is then fed into a mill which turns it into a powder.

Coal mills are of differing types. The three main types in use today are ball mills, bowl mills, and hammer mills. A ball mill consists of a rotating cylinder or tube containing steel balls. Coal is introduced into the tube and the tumbling steel balls crush it to powder. A bowl mill uses heavy rollers which run against a steel ring or bowl. As the bowl and rollers rotate against one another, coal introduced into the bowl is crushed between the bowl and the rollers. A hammer mill uses hinged hammers to crush the coal against a steel striking plate. In all types pneumatic transportation (air) is used to transport powdered coal through the mill and on to the burners. This pneumatic transport also allows classification of the powder so that large particles are returned to the mill to be reground while heavy impurities such as iron pyrites are separated from the coal. Typically a mill will be controlled so that 70% of the coal powder it produces will pass through a 200 mesh, indicating a particle size of less than 75 microns. Operation of coal mills is affected by both the moisture content and volatiles content of the coal and so will vary from coal to coal, therefore coal processing must normally be tuned to a particular type of coal.

The powdered coal exiting the mill is carried pneumatically directly to a burner in the combustion chamber. The air carrying the coal forms the primary air for the combustion of the fuel in the boiler furnace.

The increasing sophistication of control and automation systems in power stations has led to many of these coal-handling processes from unloading of coal into the stockyard right through to the delivery of powdered coal to the plant boiler and combustion chamber to be automated.

THE DEVELOPMENT OF THE POWER PLANT BOILER

The boiler in a coal-fired power plant (more accurately the combination of the furnace and boiler, but usually referred to simply as the boiler) is the site of one of the two key energy conversion processes that take place in any combustion power plant. The boiler's role is to release the chemical energy contained in the coal and convert it into heat energy. The heat energy is then carried away as steam produced when the heat is absorbed by water. This heat energy is subsequently converted into electrical energy in the second energy conversion system, the turbine.

The early boilers that were used to supply steam for steam engines were not much more than large kettles. Sometimes called haystack boilers because of their shape, these could provided a large volume of steam but the pressure of the steam was little more than atmospheric pressure. These were soon superseded by more complex designs in which water was contained in an entirely closed, often cylindrical, vessel that allowed a pressure of steam to build up, providing "strong steam" or high-pressure steam to drive an engine. A furnace was placed at one end of this vessel and the hot gases from the furnace passed through a tube which ran down the center of the water-vessel, heating the water and producing steam in a space left at the top of the vessel. This steam was then extracted through pipework and carried to the steam engine. Later boilers extended to the path of the tube carrying the hot furnace gases back through the water (a two-pass boiler) and even back again to create a three-pass boiler and improve heat transfer to the water. These boilers with large flues carrying the combustion gases are generally called flue boilers. A later adaptation was to pass the hot gases through a bundle of small tubes instead of one large one, leading to the development of the main early forerunner of the modern boiler, the fire tube boiler. A version of this type of boiler is shown in Figure 3.2.

A further refinement in some early designs was to superheat the steam created by the main boiler by passing it in pipes through the hot gases before they exited to the stack. This created steam at a higher

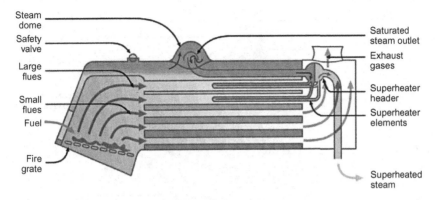

Steam dome
Safety valve
Large flues
Small flues
Fuel
Fire grate
Saturated steam outlet
Exhaust gases
Superheater header
Superheater elements
Superheated steam

Figure 3.2 A fire tube boiler. Source: Image courtesy of Wikipedia Commons.

temperature and pressure and removed droplets of water that are contained in normal steam by heating the fluid well above the boiling point of water. This, in turn, helped prevent damage to the engine components that could be caused by entrained water droplets eroding metal. Superheating also allows more heat energy to be carried by a smaller quantity of water, reducing the size of boiler necessary for a given power production.

Fire tube boilers, in which the combustion gases from the furnace pass through multiple tubes within a large cylindrical pressure vessel containing water, were used for locomotives and in many marine applications but they were limited for stationary applications because they could only achieve a pressure of around 1.7 bar. They were also liable to catastrophic failure in the event that the pressure vessel ruptured and released the pressurized water it contained.

In order to overcome this limitation a new design called a water tube boiler evolved. In this type of boiler the water was passed through bundles of tubes inside the furnace. These tubes ran between a feed-water drum containing cold water at the bottom of the water-steam system and a hot steam drum at the top. The water circulated from the feed-water drum, through the water tubes and then into the steam drum, absorbing heat as it passed through the furnace and eventually generating steam. Steam was then drawn from the top of the steam drum and could be further superheated before being taken to the steam engine or turbine. Steam exiting the turbine was condensed and returned to the feed-water drum. An example of this type of boiler is shown in Figure 3.3.

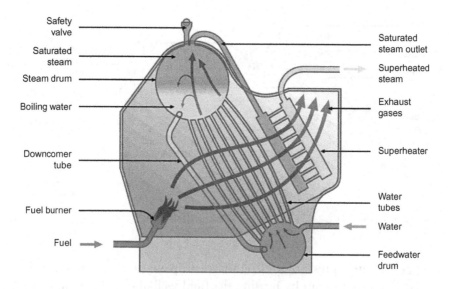

Figure 3.3 A water tube boiler. Source: Image courtesy of Wikipedia Commons.

At first these new water tube boilers used brick furnace enclosures. These designs could generally only produce up to 23,000 kg/h of steam, limiting the size of the power plant they could support. As a consequence many small electricity utility companies grew up in towns and cities, each supplying a small area from these small boiler/turbine systems.

All-brick furnace designs were limited by their structural strength and by the temperatures they could support. The brick walls were up to 56 cm thick in order to retain all the heat generated and as sizes increased absolute size became a limitation. This led to a new design where instead of the water tube bundles being placed in the center of the furnace, the tubes were constructed in the walls. By this means the tubes containing the water would cool the walls of the furnace, controlling heat loss so that thickness could be reduced massively. Lightweight external insulation was then added to reduce residual heat loss further.

The first of this new type of design, called a tube-and-tile furnace, had water tubes fitted into specially shaped fire bricks which formed the furnace wall. This was soon superseded by designs in which the tubes themselves sometimes joined by fins, sometimes joined tangentially, tube to tube, formed the actual internal wall of the furnace. This was the foundation for the modern water wall boiler design. In most recent boilers the boiler is constructed from a membrane water wall in which vertical tubes are linked together by fins in sheets to create the furnace enclosure. Behind the water wall is an insulation layer, which is then capped with external steel lagging.

The development of the water wall design allowed much bigger boilers to be built. Meanwhile the development of the radiant boiler, which had only one drum instead of two, also helped increase capacity. Meanwhile coal preparation and furnace design advanced too. The first pulverized coal boiler was built in 1918, and by the 1930s single boilers capable of delivering 450,000 kg/h of steam were being manufactured. Modern manufacturing techniques developed during the 1950s and 1960s pushed the maximum output of a single boiler to around 1,800,000 kg/h. Since then advances in materials and design techniques have allowed the largest modern boilers to achieve 4,000,000 kg/h or more in a 1300 MW power plant.

MODERN BOILER DESIGN

Large pulverized coal plant boilers use water wall construction techniques to build the furnace enclosure in which the coal is burnt. Multiple burners sit in openings in the sides of the boiler enclosure and these feed a mixture of powdered coal and air into the boiler combustion chamber. Combustion takes place under tightly controlled conditions with the aim of creating a fireball in the center of the combustion chamber where the pulverized coal particles burn rapidly, releasing radiant heat which is captured at the walls.[2] There is some convective heat transfer too.

Using this method of construction, single pulverized coal boilers have been built to provide steam for steam turbine generators of between 50 MW and up to 1300 MW in capacity. For optimum efficiency a plant needs to be of at least 300 MW but single units capable of more than 700 MW are rarely built because of the effect on an electricity supply system if a single unit of this size or greater should suddenly have to be taken out of service. A simplified cross-section of a large coal plant boiler is shown in Figure 3.4, illustrating the steam and water circuit.

---

[2]There is normally a facility within the plant to use inject light oil or gas into the plant to initiate combustion within the furnace since powdered coal is difficult to ignite on its own.

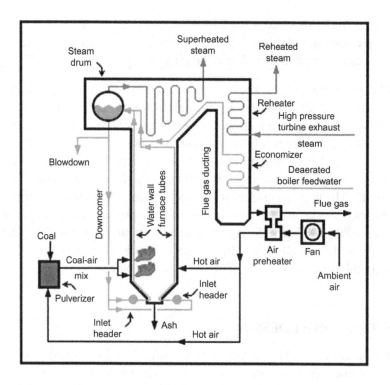

Figure 3.4 Cross-section of a power plant boiler. Source: Image courtesy of Wikipedia Commons.

The burners that feed the fuel and air into the combustion chamber are fed from the coal processing systems. The pulverized coal is carried into the combustion chamber from the coal mills using high-temperature air which has been preheated using residual heat from either the boiler or boiler-feed water. This air is called the primary air because it is mixed directly with the coal. The composition of the air/pulverized coal mixture has to be carefully controlled to ensure that combustion takes place under specific chemical conditions. Parameters such as the flame temperature and speed of combustion will depend on factors such as the calorific content of the coal, the amount of ash it contains, and the proportion of volatile material within the coal. As these vary, so the coal-to-air composition must be varied.

The primary air does not normally provide sufficient oxygen for combustion of all the coal in the combustion chamber so the burner nozzle will also be provided with a supply of secondary air, the volume of which can be adjusted to ensure the correct combustion conditions as the coal/air mixture enters the boiler and ignites. (Powdered coal will not ignite easily so when furnaces are started up either oil or gas will usually be used to initiate ignition.)

The number of burners that provide fuel for the combustion chamber will depend on the size of the boiler. Their arrangement can also vary depending upon the design of the unit. Some boilers use burners which are mounted on one side of the chamber only. A second option is to have opposing burners, on facing sides of the boiler. A third option is to have boilers in each corner of the combustion chamber; these are often referred to as tangential burners because they fire their coal and air at a tangent to the fireball in the center of the furnace. In most cases there are no obvious advantages to the different arrangements and it tends to be a matter of the preference of the manufacturer. However, multiple burners do appear to help to create a stable fireball within the furnace.

The temperature of the fireball in the combustion chamber is between 1500−1700°C in the hottest part of the flame (in some cases it can exceed 1800°C) for bituminous coals. For coals with lower calorific content the temperature is likely to be in the region of 1300−1600°C. Under the conditions in the boiler most the ash that remains after combustion is in the form of small particles which are carried away with the combustion gases. These particles are then collected by a gas cleanup system. Boilers of this type are called dry-bottom boilers because there is little or no ash or slag to collect from the bottom of the boiler. Control of the fireball so that it remains in the center of the combustion chamber will also prevent fouling of the water walls that make the furnace enclosure.

The high combustion temperature within the combustion chamber means that nitrogen in coal and air can easily be oxidized to generate nitrogen oxides. In order to prevent this, combustion takes place in stages. During the first stage of combustion the amount of oxygen available is restricted, creating reducing conditions under which most of the oxygen available is captured by carbon during the combustion reaction to create carbon dioxide and none remains to react with nitrogen. As the combustion gases begin to cool, tertiary air is added through ducts higher up within the combustion chamber. The air is sufficient to allow the combustion of carbon to go to completion, while the slightly lower temperature reduces the risk of nitrogen oxide production. Primary combustion can take place in much less than one second. Total particle residence time within the boiler is of the order of 2–5 s. During this time over 99% of the carbon should have been consumed for maximum efficiency.

Actual overall efficiency will depend on the boiler design and on the type of coal. The best plants burning bituminous coals can achieve 45% to 47% net efficiency. However, similar plants burning poorer-quality brown coal or lignite may only achieve 42%. Ambient temperature at the plant site can also affect overall efficiency because it controls the lowest temperature that can be achieved when condensing steam at the exit of the steam turbine.[3]

There are two overall boiler configurations that are popular for modern pulverized coal-fired power plants, tower boilers and more conventional two-pass boilers. A two-pass boiler consists of the primary boiler and combustion chamber constructed from water walls responsible for the main heat absorption and some water-tube bundles at the top of the enclosure to control the furnace-exit-gas-temperature. The combustion gases then turn through 180° and flow down through an extension of the boiler chamber where additional heat transfer sections are located as tube bundles within the flue gas path. The alternative tower boiler does not turn through 180°. Instead all the heat transfer elements are mounted vertically, one above the other over the combustion chamber. This type of design has become popular in Europe in recent years.

The additional heat transfer elements within both types of boiler include superheaters which are used to heat the steam well above the boiling point of water, economizers that are used to preheat boiler water when it is fed back to the boiler from the condenser (economizers may also be used to preheat combustion air), and reheaters which add additional heat to steam between turbine elements of the station's steam turbine.

SUB-CRITICAL AND SUPERCRITICAL BOILERS

The heat released during coal combustion is partly radiant heat and partly convective heat. Radiant heat is captured by water running in tubes within the walls of the combustion chamber. Further collections of tubes are placed in the path of the flue gases exiting the combustion chamber and as water passes from one set of tubes to another its

---

[3]This will affect the overall temperature difference between steam turbine entrance and exit, a parameter that controls overall Carnot cycle efficiency.

temperature rises and finally steam is generated. In a conventional boiler this will take place within a steam drum which allows the phase change between liquid and gas to proceed smoothly. Steam from the drum may them be superheated to create an even higher-temperature gas.

The efficiency of a steam plant depends on the temperature and pressure of the steam that is delivered to the steam turbine. Improved efficiency can be achieved by increasing both. However, there is a point, as both these increase, at which the nature of water changes and the distinction between the gaseous and the liquid states ceases to exist. Water in this state is said to be super-critical because it has passed the supercritical point in the water/steam phase diagram. The supercritical point for water occurs at 22.1 MPa/374.1°C.

Plants that operate with steam conditions below the critical point are referred to as sub-critical plants and these dominated coal plant construction during the last 50 years of the twentieth century. However, many modern pulverized coal plants now operate with supercritical boilers. Such units require special materials in order to withstand the high temperatures and pressures to which they are exposed. On the other hand, operation under supercritical conditions simplifies the overall steam system configuration because a steam drum is no longer required. Since the gaseous and liquid phases are no longer distinguishable, water can be converted smoothly into steam within the boiler tubing without the need for the drum.

Steam conditions at the exit of a modern sub-critical power plant boiler vary widely but typical figures would be a steam temperature of 540°C and a steam pressure of 17 MPa. Sub-critical plants burning high-quality coals are generally capable of generating electricity with an efficiency of 35–36% though some may reach 38%, depending upon site conditions. Supercritical power plants operate at steam exit temperatures of 540–600°C and at exit pressures of 23–30 MPa. These plants can achieve efficiencies of up to 41%. There is a further sub-division called ultra-supercritical power plants which operate at even higher temperatures and pressures. Although the definition of an ultra-supercritical boiler is not precise,[4] plants operating at these more

---

[4]The US Electric Power Research Unit has defined an ultra-supercritical power plant as one operating with a boiler exit steam temperature above 593°C.

extreme conditions have been capable of up to 47% efficiency under optimum site conditions. In comparison, the average efficiency of coal-fired plants operating across the globe is around 28% and that of the US coal fleet is around 33%.

Ultra-supercritical boilers are not new. Plants were built during the 1950s that operated with steam temperatures of 650°C and a steam pressure of 35 MPa. However, the lack of suitable materials made these plants difficult and expensive to build.

The high temperatures and pressures in supercritical boilers make extreme demands on the materials used to construct them. While steam boilers have traditionally been constructed from steels, conditions in ultra-supercritical power plants require that some components are made from nickel-based alloys similar to those used to construct the high-temperature components used in gas turbines. These materials are more costly than steels.

With further materials development it is expected that future plants will be able to operate at steam temperatures of 700–750°C. This should permit an energy to electrical conversion efficiency as high as 55% to be achieved in a coal-fired steam plant.

While efficiency is the most important factor driving boiler design, flexibility has also been recognized as vital in recent years. Coal-fired power plants have traditionally operated as base load power stations operating essentially at full output all the time. This is no longer the situation everywhere. In some regions coal-fired power stations are being used to support the generation of renewable electricity. This means they have to be able to operate both efficiently and effectively at part load as well as full load and to be able to change output as required by the grid. One technique being used to achieve this is sliding pressure operation under which the steam pressure is allowed to fall as output falls but steam temperature is maintained. With sliding load operation it is possible to maintain relatively high efficiency at part load, even though this may involve falling below the critical point of water.

The construction of the boiler plant also has a major effect on its flexibility. Boiler water walls need to be constructed from thick, strong materials to withstand the most difficult conditions experienced in ultra-supercritical plants. Compared to thinner steels, these will take much longer to heat up or cool down and this thermal inertia can affect the speed at which they are capable of changing their operating conditions. A coal-fired power plant built for flexibility may therefore have to use thinner materials and operate under less extreme conditions, compromising the ultimate level of efficiency it can achieve.

Increasing the steam conditions, allowing greater energy efficiency from the Carnot-cycle controlled steam turbine, is the most important way of increasing plant efficiency but other small measures can also help. For example, capturing and reusing as much heat in the exhaust gases before releasing them to the atmosphere and reducing the condensing pressure at the exit of the steam turbine can also add to overall efficiency, as can using more than one reheat stage between steam turbines in a plant with multiple turbines.

CHAPTER 4

Steam Turbines and Generators

When the electricity-generating industry began, the only type of steam engine available to drive a generator was a reciprocating or piston engine. These had reached a high level of development by the end of the nineteenth century but even the best of them was restricted as to the absolute mechanical power output it could produce. Physical restrictions on weight and size meant that larger machines were impractical.

The principle upon which the reciprocating steam engine operated was the use of the potential energy in high-pressure steam to drive a piston in a cylinder, venting the exhausted steam to the air via a valve while the inertia of the engine returned the piston to its starting position and the cycle restarted. Charles Parsons, inventor of the steam turbine, sought, instead, to use the velocity and kinetic energy of a jet of steam released from a high-pressure source to drive an engine. It was already well known that a jet of fluid could be used to drive a wheel by mounting blades upon the wheel, upon which the jet impinged. This is the basis for the simple waterwheel. The Greek engineer, Hero of Alexandria, had also demonstrated that allowing high-pressure steam to escape from nozzles mounted tangentially on a boiler would drive the boiler to rotate. His *aelopile*, sometimes called Hero's engine, is the first example of a reaction engine as used in a rocket.

The problem with harnessing this principle to build a steam engine in which the steam drove a bladed wheel was the speed of the jet of steam. Even a moderately low-pressure steam jet can reach 750 m/s and high-pressure steam exiting a nozzle into a vacuum might reach 1500 m/s. In order to harness this efficiently to capture energy using a bladed wheel, the blade has to be traveling at around half the speed of the steam. This would require the wheel to rotate at an extremely high speed. Aside from the basic engineering problems this raised there was the question of the materials being able to withstand the centrifugal force to which they would be exposed.

Coal-Fired Generation. DOI: http://dx.doi.org/10.1016/B978-0-12-804006-5.00010-1

The solution that Parsons adopted to this problem was to reduce the speed that the steam reached during expansion by breaking the expansion into a series of small stages. He achieved this by building an engine that comprised a rotor with a series of blades that fitted into a cylinder that also had blades fixed inside it. The rotor blades fitted into the casing between the fixed blades. When steam from a boiler was released into the front of the engine it would pass alternately between the fixed and moving blades. The fixed blades acted as nozzles through which the steam partially expanded and the velocity it acquired was then transferred to the next set of blades which were moveable, causing them to rotate. At the same time, passages between the moving blades also acted as nozzles and as the steam expanded through these, the moving blades were further impelled by the expansion, a reaction in the manner of the engine of Hero of Alexandria.

By using multiple stages of fixed and rotating blades, the expansion of the steam was controlled to take place in small increments and the absolute velocity of the steam in each stage was made manageable. This allowed an efficient steam turbine to be built. However, even this rotated at around 300 rev/s or 18,000 rev/m, much faster than the 1500 rev/m typical of a reciprocating steam engine of the day. As a consequence Parsons had to design a dynamo that could be driven by his new engine. The first of these, now at the Science Museum in London, produced 75 Amps at 100 V, an output of 7500 W.

A Parsons turbine utilizes the *action* of high-speed steam on the rotating blades and also the *reaction* on the blade of the steam expanding through the nozzle created by the moving blades. It is also possible to build a turbine that exploits solely the action of the high-speed steam, creating what is known as an impulse turbine. However, these are generally less efficient than a turbine that exploits both the impulse and the reaction effect.

The Parsons turbine were initially used for ships' lighting. It required Parsons himself to build power stations, first in Newcastle upon Tyne in 1890, then Cambridge in 1892 and Scarborough in 1893, for the new technology to start to be widely adopted for land-based generation. By 1900 there were 1000 kW turbine generators available, and in 1912 the first 25,000 kW machine was built. Today single-rotor turbines with outputs of 250 MW are common. However, large power plants with generating capacities of 1000 MW to 2000 MW are generally equipped with multiple turbines to extract the maximum amount of energy most efficiently.

Steam turbines for large coal-fired power stations are generally broken down into three sets, high-pressure turbines, intermediate-pressure turbines, and low-pressure turbines. The high-pressure turbines are the smallest of the set, with the shortest blades and they receive steam from the boiler at the highest temperature and pressure. In a basic layout the steam from the high-pressure turbine then enters the intermediate-pressure turbine. This has longer blades and is optimized for steam at a lower entry temperature and pressure. The high-pressure and intermediate-pressure turbines are usually on a single shaft, driving a single generator.

The low-pressure turbine or turbines – as there may be more than one of these – may also be mounted on the same shaft as the high- and intermediate-pressure turbines. However, it is often preferable to operate these turbines at a low rotational speed as they have very long blades and the tip speeds could otherwise reach velocities beyond the limits of the materials from which they are made to withstand. Steam exiting the low-pressure turbines is condensed at as low a pressure as can be achieved in order to gain the maximum energy from the steam. Heat may be recovered from the condensed water to heat combustion air before the water is returned to the boiler and recycled. A 3D section of a typical large power plant steam turbine with three sections is shown in Figure 4.1.

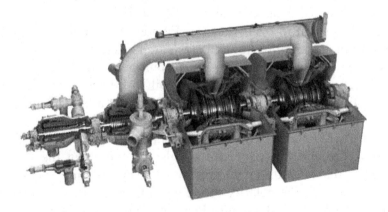

Figure 4.1 Cross-section of a large steam turbine with high-pressure, intermediate-pressure, and two low-pressure sections. Source: Image courtesy of Siemens.

A more complex refinement to steam turbine layout involves steam reheat between turbine stages. The most common way of implementing this is to take the steam from the high-pressure turbine and return it to the boiler where its temperature is raised again before it enters the intermediate-pressure turbine. This can convey a significant efficiency advantage. One reheat stage is common. Less common is a further reheat stage between intermediate- and low-pressure turbines. This can increase efficiency further, but at a cost.

The best modern coal-fired power plants with ultra-supercritical boilers can come close to 47% efficiency. If steam temperature is pushed to 700°C it should be possible to reach 50% efficiency and at 750°C it may be possible to achieve 55% efficiency. However, these temperatures push existing materials to the extremes of their capabilities and new, more expensive alloys are needed to be able to build plants capable of operating under these conditions.

GENERATORS

The last stage in the energy conversion process that starts with coal and ends with electricity is the generator. This electro-mechanical machine converts the rotary motion produced by the power station turbine into electrical energy that can be delivered to the transmission and distribution system.

All modern generators are based on a phenomenon discovered by Michael Faraday at the beginning of the nineteenth century. If a conductor is passed through a magnetic field, this movement generates an electric current in the conductor. Faraday built a very simple electromagnetic generator, now called Faraday's disk, which demonstrated the principle of continuous current generation. However, the first effective electric generator was built by the French instrument maker Hippolyte Pixii. His machine consisted of a permanent magnet which was turned by hand. The north and south poles of the magnet alternately passed a coil of insulating wire wrapped around an iron core and this produced a pulse of current in the coil. The pulses alternated in different directions but by designing a commutator that reversed the polarity of the wires from the coil where it connected to an external circuit, Pixii was able to produce a direct current output.

Other scientists experimented with rotating coils within the poles of a permanent magnet. However, all these systems produced a pulsed DC current with low average output. By replacing the two-pole rotating coil with a multi-pole coil and a series of synchronized commutators it became possible to produce a better, more continuous current.

Machines with permanent magnets were called magneto-electric machines. However, permanent magnets limited the capability of these early generators. When the permanent magnet was replaced by an electromagnet, with a magnetic field generated from an external (exciting) direct current source, much more powerful magnetic fields could be produced. Machines using these were initially called dynamo-electric machines but this was eventually shortened to dynamo.

The next important discovery was made independently by the Englishman Henry Wilde and the German Werner von Siemens. Both men discovered a way of allowing the dynamo to provide the electric current from a secondary generator to power the magnetic field of the electro magnet, allowing these machines to be self-exciting. Doing away with a separate electrical source for the excitation current enabled much larger dynamos to be built.

Early generators provided a direct current because that was what was required for the local lighting circuits that they supplied. As the industry grew, and with it the need to transport the electric power over longer distances, the advantages of an alternating current supply became apparent. The voltage of an alternating current supply can be raised or lowered easily using a transformer, something that is not possible with a direct current. For long-distance transport of electricity it is more efficient to use a high voltage and a low current because resistive losses are proportional to the size of the current flowing in a cable. This high-voltage supply can then be reduced to a lower voltage/higher current close to the point of use. This high-voltage transmission/low-voltage distribution forms the basis of the modern supply system topology with its separate transmission and distribution systems.

Alternating current generators, which did away with commutators, were called alternators and these became the standard generators for all power stations in the early years of the twentieth century. The common form of a large power station generator is to have a rotating magnetic field generated by a rotor which turns inside a fixed coil, called the stator. Most modern electricity supply systems have three AC phases at 120° to each other. These are generated by building three independent coils into the stator, each occupying one third of the circumference.

The rotor of a large generator can contain a single two-pole coil producing one north and one south pole that rotates within the rotor. However, depending upon the speed at which the rotor turns, there may be more poles in order to produce an output at the required frequency. The faster the rotor turns, the fewer poles are required. A rotor that turns at 3000 rev/m (50 rev/s) requires only two poles to be able to generate an AC current at 50 Hz. (For a 60 Hz system the equivalent would be 3600 rev/m.) This would be suitable for the high-pressure and intermediate–pressure turbines in a large coal-fired power plant. However, this rotational speed might be too high for the much larger low-pressure turbines and these may operate at half or one quarter of the higher speed. At 1500 rev/m the rotor would need four poles to provide a 50 Hz output and for 750 Hz the number required is eight poles.

Early generators were relatively small and problems such as heat dissipation were not a major issue. However, as the size of generators grew, so did the size of the heat problem. A modern generator can achieve an efficiency of more than 98% and some exceed 99%. However, with generators of hundreds or even more than 1000 megawatts, the amount of heat that must be dissipated is of the order of several megawatts.

Cooling both the rotor and the stator is the key problem for designers of large generators. There are three coolants in general use, air, hydrogen, and water. Air is the cheapest cooling to implement but it is the least effective of the three for carrying away heat. Modern manufacturers can rely on air cooling for generators of up to a maximum of around 350 MW. By optimizing the cooling flow with the generator, manufacturers expect to be able to push this to perhaps 400 MW in the near future.

Hydrogen is around three to four times as effective at cooling a generator as air. However, the gas is explosive so must be controlled carefully. Hydrogen-cooled generators are normally operated at a hydrogen pressure of 2–3 bar to improve the cooling capabilities and to ensure that air cannot enter the generator enclosure and produce an explosive mixture. However, leaks have to be carefully monitored. In addition,

Figure 4.2 Cross-section of a hydrogen-cooled generator. Source: Image courtesy of Siemens.

hydrogen requires a gas cycling system and an external heat exchanger to cool the hydrogen that has passed through the generator. A 3D cross-section of a hydrogen-cooled generator is shown in Figure 4.2.

For the very largest generators, water cooling is necessary. This is 50 times more effective than air but is much more complex to implement than either air or hydrogen cooling. Generally water cooling is only used for the stator of a large generator while hydrogen cooling is employed for the rotor.

The efficiency of power plant generators is already very high so making significant gains is difficult. One area where gains can be made is by reducing the magnetic losses within the cores of the rotor and stator and the electrical losses in the coils themselves. This can be achieved with higher-quality materials. However, these materials are significantly more expensive and may not be cost-effective. One radical change that may eventually become possible is the use of high-temperature superconducting materials for coils, thereby reducing coil losses significantly. This may add a few tenths of a percentage point to overall efficiency but again cost will be a major consideration.

A new consideration for coal plant generators is the need to be able to operate at variable output in order to support renewable generation on a system grid. Cycling the output of a generator can cause significant aging and wear problems and redesign of some components may be necessary to accommodate this type of duty cycle.

CHAPTER 5

Fluidized Bed Combustion and Coal Gasification

The pulverized coal power plant is the most efficient large-scale coal-based generating plant but it is most effective at a very large scale with units of up to 700 MW in size and single plant capacities of 2000 MW and more. There are alternatives which can be more cost-effective at smaller scale. One of the main alternatives is the fluidized bed combustion plant.

This type of plant can burn a wider range of fuels than a pulverized coal plant, fuels including poor-quality coals, waste materials, and biomass, as well as high-quality coals. It operates at a lower temperature than a pulverized coal plant, with a typical combustion temperature of 1000°C, so the generation of nitrogen oxides is much less of a problem. In addition, it is possible to incorporate sulfur capture within the fluidized bed itself rather than in a separate chemical separation plant that scrubs the exhaust gases, a significant simplification of the overall plant. Further, fluidized bed combustion provides a more uniform temperature in the combustion zone, making heat capture and transfer possible with smaller capture surfaces and this can, in principle, reduce the overall size and cost for a given heat output.

The principle of the fluidized bed is simple and elegant. The fluidized bed is a bed or layer of small solid particles, normally with diameters of less than 6 mm in the case of a power plant — although they can be larger in other applications — held initially at the bottom of a combustion chamber. In operation, air is blown into the chamber through ports in its floor. This air has sufficient velocity that as it passes up through the particles it entrains them so that they become suspended above the bottom of the combustion chamber floor. How far they are suspended above the floor depends on the air velocity. This suspension of solid particles now displays many of the properties of a fluid, so that the particles move and jostle one another as would molecules in a liquid or gas. The small particles also have a large

Coal-Fired Generation. DOI: http://dx.doi.org/10.1016/B978-0-12-804006-5.00011-3

surface area. These two factors allow solid state reactions between particles of different reactants to take place quickly as well as reactions between the solid particles and the gas phase in which they are suspended. Both aspects are important for the operation of fluidized beds.

The person normally credited with the construction of the first fluidized bed is the German chemist Fritz Winkler. While working for the company BASF, Winkler was trying to improve on the production of synthesis gas (a mixture of mainly hydrogen and carbon dioxide produced from carbon-containing fuels such as coal) for the production of ammonia. In 1924 he developed the principle of the fluidized bed technique and applied it to the coking of fine-grain lignite, generating an excellent combustible gas according to reports from the time.[1]

The technique was not initially applied to energy production but it was used in a range of industrial processes. As well as synthesis gas production, BASF developed it as a means of roasting pyrites in order to make sulfuric acid and it was used for other chemical and metallurgical processes. Today the fluidized bed technique is used in a wide range of industrial and commercial applications, including fast freezing in the food industry for small items such as peas.

The fluidized bed developed by Winkler is what is now commonly known as a bubbling fluidized bed (BFB). Work on this technology was taken up in the USSR in 1940 and in China in the 1960s, and by UK and US organizations during the 1970s when it was applied to power generation. Meanwhile at the end of the 1930s two scientists at the Massachusetts Institute of Technology, Warren Lewis and Edwin Gilliland, began to develop what has now become known as the circulating fluidized bed (CFB). Initially intended for mineral oil cracking, this design was later adapted for power generation applications too and has since become the main type of fluidized bed in use for coal combustion. Meanwhile, during the 1980s a refinement of the BFB called a pressurized fluidized bed combustion (PFBC) reactor was developed as a more efficient version of the standard bubbling bed design. Pressurized versions of a CFB have also been proposed though none has yet been built. Today all three designs are exploited in different parts of the power industry.

---

[1]Winker's contribution is documented in BASF's historical archives.

THE BFB REACTOR

The BFB is the simplest of the fluidized bed reactors that has been developed for power generation applications. The basis of the operation is similar to that of a pulverized coal boiler with a combustion chamber constructed using water walls and further heat capture surfaces in the hot gas path before the flue gases exit the boiler. However, the combustion process takes place in a bed at the bottom of the combustion chamber instead of in a fire ball suspended within it and further heat-capture water-tubes are often placed within the BFB itself, leading to very efficient energy capture.

The bed is composed primarily of a refractory and incombustible material, often sand, to which coal particles are added via feeders located above or to the side of the bed. In a typical BFB, only 5% or less of the bed is coal while the remainder is the inert bed material. In order to keep the bed in a liquidized state, high-pressure air (the primary combustion air) is introduced through the bottom of the bed and there may also be further air inlets in the sides of the chamber providing secondary combustion air. Air feeds are normally preheated using residual heat from the plant.

In coal combustion, the reaction taking place within the bed is a solid phase—gas phase reaction between coal and air. This will take place rapidly under the conditions within the bed, although reaction time is generally longer than in a pulverized coal furnace because of the lower temperature. In addition, the addition of limestone to the refractory bed material enables a similar gas—solid phase reaction to take place between any sulfur dioxide generated during combustion from sulfur in the coal and the limestone particles, capturing and removing the sulfur dioxide from the flue gases.

Since the temperature in the fluidized bed it typically around 950°C instead of up to 1700°C in a pulverized coal combustion chamber, the rate at which nitrogen oxides are formed from air is much reduced. This may remove the need for additional nitrogen oxide removal, depending upon the environmental restrictions in force at the power plant site. Even where further capture is necessary, the quantities and concentrations involved will be relatively lower than for a high-temperature combustion plant.

The BFB was developed as a coal-burning technology up until the 1980s. Units were built in the USA and extensively in China, where by

1980 there were around 200 in use.[2] However, the technology fell out of use because the overall combustion efficiency and the efficiency of sulfur capture was judged relatively unfavorable compared to the alternative CFB technology. In addition, erosion of the water tubes within the BFB itself became an issue, as did the ability to scale up the bubbling bed to larger sizes to compete with the alternative pulverized coal technology. However, the use of BFB technology has continued with both biomass and waste material where the small scale and overall efficiency are less of an issue. In these cases there are often no heat transfer tubes within the BFB and there may be an additional pulverized coal burner above the fluidized bed combustion chamber to improve efficiency when burning waste or biomass.

THE CIRCULATING FLUIDIZED BED

Whereas the air blown into a BFB creates an easily defined suspended layer of solids, the air blown into a CFB is at much higher velocity and when it picks up and entrains the particles, it spreads them much further vertically, in some cases carrying them up to the top of the combustion chamber. Since these entrained particles must not escape the combustion plant, in a modern CFB plant the flue gases and entrained particles pass from the top of the combustion chamber into a cyclone filter which captures the solid particles and returns them to the bottom of the combustion chamber while allowing the flue gases to exit the boiler. Initial particle size is less than 10 mm for coal, under 50 mm for biomass, and smaller than 1 mm for limestone used to capture sulfur. The coal particle size is much larger than in a pulverized coal plant and can be achieved with a coal crusher instead of high-performance coal mills. A schematic of a CFB power plant is shown in Figure 5.1.

The advantage of the high-velocity fluidization in the CFB is that it accelerates the reaction between the solid and gaseous phases, a feature that was first recognized by Lewis and Galliland in the 1940s. However, it was not until the 1970s that a dedicated coal combustion plant based on this design was first constructed. Since then a number of designs have been proposed, each with slightly different features. One of the key variables is the velocity of the air that enters from the

---

[2]*Features of Development of BFB-CFB Combustion*, Bo Leckner, 69th IEA-FBC Technical Meeting, Aix-en-Provence, September 2014.

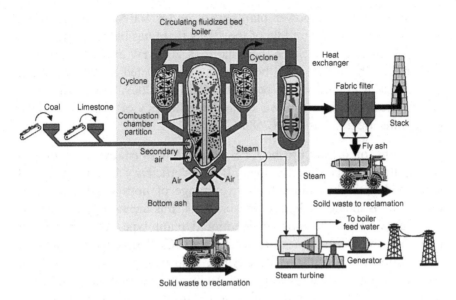

Figure 5.1 Cross-section of a circulating fluidized bed boiler power plant. Source: Image courtesy of the US Department of Energy.

bottom of the combustion chamber. The fluid nature of the fluidized bed leads to a spread of particles depending upon their density, with more dense particles remaining lower in the chamber while less dense particles are carried higher. In some CFB designs the air velocity is relatively low and there is a pronounced bed at the bottom of the combustion chamber and a much lower particle density towards the top. This is sometimes called a turbulent bed to distinguish it from the BFB and the CFB. In fast fluidized beds, on the other hand, the bed is spread vertically to the top of the combustion chamber and there is no identifiable bed towards the bottom.[3]

The CFB can remove around 90–95% of the sulfur contained in the coal it burns, compared to the BFB which can only achieve 70–90% removal efficiency. Energy conversion efficiency of around 43% is achievable, close to that of a pulverized coal plant. However, this level of efficiency can normally only be reached with large CFB plants that employ supercritical steam conditions and large, efficient steam turbines.

[3]In reality a fluidized bed will probably contain elements of each type with larger, more dense particles forming a bubbling bed at the bottom of the chamber while lighter particles are carried away.

The earliest CFB power plants were under 100 MW in generating capacity but since the middle of the 1990s there has been a major effort to build larger plants. Since then plants of 200 MW and 300 MW have been built. The largest CFB boiler so far constructed is a 460 MW unit at Lagisza in Poland which began operating in 2009. This plant uses a supercritical boiler to achieve a claimed efficiency of over 43%. Steam conditions are 275 bar/560°C/580°C.

The aim of scaling up the CFB unit size is to enable it to compete with pulverized coal technology. If high efficiency can be achieved, combined with the ability to burn a range of both good- and poor-quality fuels including lignite and biomass, this technology could offer an alternative to the more conventional plant type in the future. In addition, some companies are working on CFB designs in which the combustion air is replaced by oxygen, a strategy that can be used to reduce carbon dioxide production. Again this offers a potential alternative to a pulverized coal combustion plant with carbon capture.

PRESSURIZED FLUIDIZED BED COMBUSTION

PFBC was developed during the late 1980s and the first demonstration plants using the technology began operating during the 1990s. These plants have all been based on BFB technology. Operating the BFB at elevated pressure leads to an efficiency increase and to a more compact design. All the PFBC plants that have been built are of the bubbling bed design. A pressurized version of the CFB has also been proposed, although no units have yet been built.

The PFBC differs in one important aspect from the atmospheric BFB; it is a combined cycle plant that uses both gas and steam turbines. Operation at high pressure means that the hot flue gases exiting the combustion chamber, once cleaned, can be used to drive a gas turbine at the same time as steam produced in the pressurized boiler is exploited in a steam turbine. Heat remaining in the exhaust gases once they have exited the gas turbine is also captured to raise further steam. It is the combination of the two types of turbine which provides the route to higher efficiency. A diagram of a PFBC is shown in Figure 5.2.

PFBC plants generally operate at 1 MPa to 1.5 MPa (10–15 atmospheres) although pressures in the range 0.5 MPa to 2 MPa have been

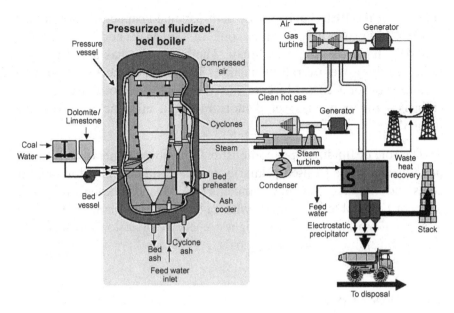

Figure 5.2 Cross-section of a pressurized fluidized bed combustion (PFBC) power plant. Source: Image courtesy of the US Department of Energy.

used. The units are particularly useful when burning high ash coals since additional refractory material simply remains in the bed. The combustion temperature within the bed is between 800°C and 900°C so nitrogen oxide production is low. Both the combustion chamber and the initial gas cleanup cyclones needed to prepare the gases for the gas turbine are contained within the pressure vessel so fuel and any sorbent added to remove sulfur must be pumped across the pressure boundary. Ash must be removed across the pressure barrier too. However, a simplification when burning good-quality hard coal is to turn the coal and sorbent into a paste, with 25% water, and then feed this mixture into the combustion chamber.

The power production from the generating units in a PFBC is broadly 20% from the gas turbine and 80% from the steam turbine. The gas turbine has to be specially designed to operate at a much lower gas inlet temperature than would be normal when burning natural gas and great care must be taken to ensure that the flue gases driving the gas turbine are free from particles and corrosive vapors that might damage the blades.

Most of the PFBC units so far built have had capacities of less than 100 MW but two larger units, one of 250 MW and a second of 360 MW, have been built in Japan. The latter also has a supercritical boiler to improve overall efficiency. The best efficiency of current designs is around 40%. More advanced designs will aim to exceed 45% efficiency but none has yet demonstrated an efficiency approaching this level.

The role of fluidized bed combustion as a power generation technology appears to be focused primarily on the combustion of low quality coals and biomass. New developments and advanced designs may lead to them being able to compete with pulverized coal plants but for the moment these latter offer the highest energy conversion efficiency for the combustion of good quality coals.

COAL GASIFICATION

As an alternative to burning coal in air to generate heat and raise steam to power a steam turbine, it is possible to convert coal into a combustible gas. This can then be burned in a gas turbine power plant or, depending upon the composition of the gas, used to provide fuel for a fuel cell.

The production of gas from coal has a long history, and town gas, a potent mixture of hydrogen, methane and carbon monoxide, was commonly used as a domestic fuel until natural gas became widely available. One form of town gas is made simply by heating coal in the absence of air, driving off its volatile components and leaving an almost pure form of carbon called coke. The coke is used in metallurgical processes such as iron production and was originally a substitute for charcoal made from wood.

Modern gasification processes usually involve a more complex reaction than the simple heating of coal. These modern processes generally require partial combustion of the coal in a mixture of steam with air or oxygen, often followed by a further reaction with water vapor to produce a gas rich in hydrogen. The partial combustion uses some of the calorific value of the coal in order to drive the overall gasification process, so a certain amount of energy is lost in this way.

Modern interest in gasification technology is based on the potential to design an efficient coal-burning power plant around a gasifier. The coal is first turned into a combustible gas, then cleaned and the gas is burned in a gas turbine. Waste heat from both the gasifier and from the gas turbine exhaust is used to raise steam which drives a steam turbine. A plant of this type, called an integrated gasification combined cycle (IGCC) plant can achieve an efficiency of around 40%. It is also possible to design an IGCC plant in which all the carbon dioxide produced is captured instead of being released into the atmosphere. This configuration is discussed below.

There are a number of different gasifier designs that have been developed for modern gasification plants. The simplest is a fixed bed gasifier, a gasification vessel which is fed with fuel from the top while ash is removed from the bottom. Air is also injected from the bottom of the vessel − this type of gasifier is called an up-draft, or counter-current, gasifier − while the combustible gas is removed towards the top of the vessel. An alternative, down-draft gasifier employs steam−air/oxygen injection from the sides of the vessel while the product gas is removed from the bottom.

A more complex alternative to the fixed bed gasifier is a fluidized bed gasifier in which the gasification process takes place within a BFB reactor. As already noted above, this is an efficient way of promoting reactions between solid particles and gases. These gasifiers are particularly effective when the residues from the gasification process are corrosive since the fluidized bed keeps the ash and residues away from the walls of the reactor vessel.

Two other gasifiers have also been developed. An entrained flow gasifier is similar in some respects to a pulverized coal burner. A pulverized fuel, mixed with steam and oxygen, is burned rapidly in a combustion chamber. Combustion takes place at a higher temperature than in either a fixed bed or fluidized bed gasifier. Finally, a plasma gasifier uses an extremely high-temperature electric arc to gasify solid materials. While applicable to coal, this type of gasifier is being developed mainly for solid waste disposal.

While there are a range of complex processes that take place during coal gasification, the overall process can be considered to be the combustion of coal under reducing (limited oxygen) conditions in the presence of water vapor. The main reactions taking place are carbon combustion:

$$C + O_2 = CO_2$$

the partial combustion of carbon:

$$2C + O_2 = 2CO$$

and the water gas reaction:

$$C + H_2O = CO + H_2$$

The reaction is carried out with around 20% of the oxygen that would be needed for the carbon to be reacted completely. Some carbon dioxide is formed but the amount is limited. The combustion and the partial combustion reactions of carbon are exothermic and the energy they release provides the driving force for the water gas reaction. Other reactions that can take place include the methanation reaction:

$$C + 2H_2 = CH_4$$

and the Boudouard reaction:

$$C + CO_2 = 2CO$$

The result of all these reactions is to produce a gas that is primarily a mixture of carbon monoxide and hydrogen with some carbon dioxide and a small amount of methane. This gas is called synthesis gas or syngas and has been used as a feedstock for a variety of industrial processes. However, since the main components are combustible it can also be burned in a gas turbine. The gas will be produced in a mixture with nitrogen if the gasifier uses air rather than oxygen. Air gasifiers tend to produce a gas of lower calorific value than oxygen gasifiers and for modern IGCC plants, oxygen gasification is normally preferred.

The conditions in the gasification reactor are severely reducing so any sulfur present in the coal is converted into hydrogen sulfide or carbonyl sulfide. Both are readily removed from the gas mixture and can be converted either to pure sulfur or into sulfuric acid for industrial use. The reducing conditions mean that nitrogen oxide production is minimized too.

The gasification process can be taken a stage further by reacting the syngas over a catalyst with more steam. This process, called the water shift reaction, converts the carbon monoxide in the syngas into a mixture of hydrogen and carbon dioxide through the reaction:

$$CO + H_2O = CO_2 + H_2$$

Any methane present will also be converted into hydrogen and carbon dioxide. The final gas is now a mixture of carbon dioxide and hydrogen with some impurities. It is relatively simple to separate the two main components, hydrogen and carbon dioxide, allowing the latter to be stored away from the atmosphere and leaving hydrogen as the main product. Hydrogen can be burned in a gas turbine or in a fuel cell to generate electricity.

If gasification is to supply fuel to an IGCC power plant, the gas must first be scrupulously cleaned to remove any particulate material and impurities that might damage the gas turbine. In order for the gas cleaning to be energy-efficient it needs to be carried out at the high temperature at which the gas exits the gasifier. Otherwise energy is lost from the process.

Once cleaned, the gas is used to fuel a gas turbine generator which provides an electrical output. The hot exhaust gases from the exit of the gas turbine are then taken though a heat recovery steam generator which extracts the remaining usable heat to raise steam with which to drive a steam turbine. A plant of this type is more complex than most other types of coal combustion plant since it requires an oxygen plant in addition to the gasifier, gas cleanup, and gas and steam turbine generation elements. A cross-section of an IGCC plant is shown in Figure 5.3.

There are a limited number of major IGCC plants burning coal in existence in Europe, the USA, China, and Japan. Nominal power outputs range from 250 MW to 620 MW. One plant, in the Czech Republic, burns lignite, others burn a mixture of bituminous coal and petroleum coke, pulverized coal, and asphalt refinery residue. Typical efficiency is around 39–41% based on the lower heating value (calorific value) of the fuel although efficiencies of up to 46% are thought possible. However, all the plants currently in operation are essentially demonstration plants and there has been no full-scale

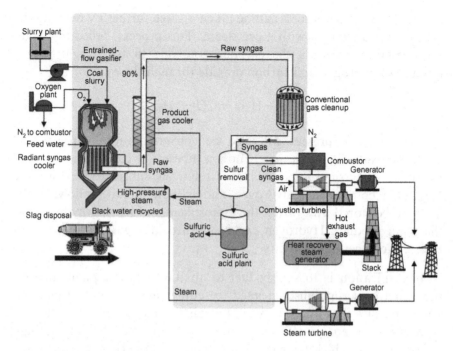

Figure 5.3 Cross-section of an integrated gasification combined cycle (IGCC) power plant. Source: Image courtesy of the US Department of Energy.

commercial roll-out of the technology. Cost remains on obstacle. Another is plant availability which appears to be lower than for more conventional plant configurations.

An IGCC plant offers one power-generating configuration based on the gasification of coal. However, if full carbon capture is being considered, leaving a gas that is rich in hydrogen, then another important option is to use the gas in a fuel cell. There are a number of fuel cell technologies commercially available and all offer relatively high efficiencies when burning hydrogen.

CHAPTER 6

Coal Combustion and the Environment

The combustion of coal is the dirtiest of the large-scale methods of generating electricity, primarily because of the range of potential pollutants that are found within the fuel. While some high-quality coals are relatively pure carbon, many are far from pure. Impurities commonly found in coal include sulfur, bound nitrogen, volatile organic compounds, heavy metals including cadmium and mercury, and a range of inert refractory materials. All of these can be released into the atmosphere during coal combustion if measures are not taken to remove them from flue gases. And then there is carbon dioxide. This gas is the inevitable product of the combustion of carbon in air and it is produced in vast quantities in electricity plants burning coal.

Most coals contain some sulfur. Often it is more than 3% of the coal and it may reach as much as 10%. When the coal is burnt this sulfur is converted into sulfur dioxide, which is carried off by the flue gases. If released into the atmosphere is can be converted into an acid. There is also organic nitrogen within coal. During combustion this is converted into nitrogen oxides of various sorts, including NO, NO_2, and N_2O. Another important source of gaseous nitrogen compounds in flue gases is the nitrogen in air which can become oxidized at the high temperatures encountered within coal furnaces. Both nitrogen oxides and sulfur dioxide can be potent pollutants.

Coal usually contains a significant amount of mineral impurity too. Some of this may melt and fuse with other similar material during the high-temperature combustion in a pulverized coal plant, creating a solid residue which is left behind in the combustion chamber as slag. This is eventually removed from the bottom of the furnace. However, depending upon the exact combustion conditions, a large proportion of the inert solid material may remain in small enough particles to be

Coal-Fired Generation. DOI: http://dx.doi.org/10.1016/B978-0-12-804006-5.00012-5

entrained and carried away with the flue gases exiting the boiler. These particles may contain heavy metals, such as cadmium and mercury which, if allowed to escape, will be released into the environment, so they too must be contained.

Some coals, particularly the bituminous varieties, contain large amounts of volatile organic compounds and these, or fragments of them generated by their incomplete combustion, can also be released. Incomplete combustion of the carbon in coal may also lead to significant levels of carbon monoxide within the flue gases. Both carbon monoxide and organic fragments can cause environmental degradation as well as affecting human health if allowed to escape.

Environmental regulations require that as far as possible these materials are removed from coal-fired power plant flue gases before the latter are released into the atmosphere. Different techniques have been developed for the most important of these; sulfur scrubbers for removing sulfur compounds, low NO_x burners and catalytic reduction systems to remove nitrogen oxides, and fabric filters and electrostatic precipitators to control dust emissions. Other trace elements, such as heavy metals, may require their own removal plants but often these can be tackled alongside one of the other pollutants, making an additional chemical treatment process unnecessary.

Air quality regulations and emission limits for pollutants from coal-fired power stations vary from region to region but most countries enforce some limits today. These tend to be strictest in the most developed countries, such as in Europe, Japan, and the USA. Table 6.1 contains figures for the concentrations of various power plant airborne pollutants that are considered permissible in the EU, and in the USA, if good air quality is to be maintained. EU regulations are generally the stricter; for example the EU expects sulfur dioxide concentrations over a 24 h period to be below 125 $\mu g/m^3$. In the USA the same standard is 365 $\mu g/m^3$. However, internationally, standards are tending to converge as the effects of even low levels of pollution on human health become more widely recognized. The PM10 particulate matter standard is for dust particles greater than 10 μm in diameter and this is generally the standard of impor-tance when considering dust from coal-fired power plants. There are other standards including PM2.5 for particles of up to 2.5 μm in diameter. For the EU the PM2.5 standard is 25 $\mu g/m^3$ averaged over 1 year.

Table 6.1 Air Quality Standards		
Pollutant	EU Standard/Averaging Period	US Standard/Averaging Period
Sulfur dioxide	125 μg/m^3/24 h	365 μg/m^3/24 h
Nitrogen oxides	40 μg/m^3/1 year	100 μg/m^3/ 1 year
Particulate matter (PM10)	40 μg/m^3/1 year	150 μg/m^3/24 h
Carbon monoxide	10 mg/m^3/8 h	10 mg/m^3/8 h
Ozone	120 μg/m^3/8 h	150 μg/m^3/8 h
Lead	0.5 μg/m^3/1 year	0.15 μg/m^3/3 months, rolling
Cadmium	5 ng/m^3/1 year	–
Source: *EU Commission, US Environmental Protection Agency.*		

Table 6.1 also shows heavy metal limits. In the EU the limit for atmospheric lead concentration is 0.5 μg/m^3 averaged over 1 year and for cadmium it is 5 ng/m^3. In the US the limit for lead is 0.15 μg/m^3 on a 3-month rolling basis. There are proposals to introduce limits on mercury emissions in the USA although these have not yet taken effect.

The figures in Table 6.1 apply to the air quality that people will encounter in the street or in their houses or offices when carrying out their daily lives. The actual emissions permitted by power plants are generally much higher than this. A power plant represents a concentrated source of pollutants but these are released in hot gases from a tall stack so that they should rise high into the atmosphere and become diluted before humans or other life-forms come into contact with them. However, the behavior of the pollutants once they enter the atmosphere is not always predictable. The behavior of the plume of exhaust gases from a power plant stack will depend on atmospheric conditions so that sometimes the pollutants will fall close to the plant, at other times they may be carried across continents.

Table 6.2 shows some of the emission levels permitted within the EU for power plant flue gases. The figures are for large plants with a thermal capacity in excess of 300 MW$_{th}$. The limits are less strict for some smaller plants. For sulfur dioxide the limit for plants built after 2003 is 200 mg/m^3, falling to 150 mg/m^3 after 2016. Permitted emission levels for nitrogen oxides are the same. Dust emissions are to be below 20 mg/m^3 after 2016 and there is a proposed emission limit for mercury of 30 μg/m^3. As earlier, these EU limits are probably some of the strictest to be found but as with air quality standards, the regulations are becoming stricter everywhere.

Table 6.2 EU Emission Limits for Large Power Plants	
Sulfur dioxide emissions for plants built after 2003	$200 \, \text{mg/m}^3$
Sulfur dioxide emission limits after 2016	$150 \, \text{mg/m}^3$
Nitrogen oxide emissions for plants built after 2003	$200 \, \text{mg/m}^3$
Nitrogen oxide limits after 2016	$150 \, \text{mg/m}^3$
Dust emission limits after 2016	$20 \, \text{mg/m}^3$
Proposed mercury emission limit	$30 \, \mu\text{g/m}^3$
Source: *EU Commission.*	

In addition to these atmospheric pollutants, there are other environmental effects associated directly with coal-fired power stations. Many plants use water for cooling and this water is often taken from lakes, rivers, or the sea and then returned, but at a higher temperature. This heating can affect local environmental conditions and life. There is noise and other forms of disruption caused by both the plant and by the delivery of coal to the plant. The latter can usually be minimized by appropriate siting of the plant.

There is one other important product of coal combustion not included in the above tables or discussion, carbon dioxide. This is the product when carbon is burnt in air, the reaction which releases the heat energy used to generate electricity.

$$C + O_2 = CO_2$$

Since the energy content of coal is virtually all contained in carbon, this fuel produces more carbon dioxide per unit of energy than any other. The flue gases from the boiler of a typical advanced coal-fired power plant may contain up to 14% carbon dioxide, although this figure will vary depending on the specific plant conditions.

The release of carbon dioxide from the combustion of fossil fuels in power plants and elsewhere into the atmosphere is widely regarded as the main cause for a steady but accelerating rise in average global temperatures over the past 150 years. Table 6.3 shows how global atmospheric carbon dioxide concentrations have increased since 1700, before the Industrial Revolution started. Levels then are estimated to have been between 270 ppm and 280 ppm. The use of coal started to accelerate at the end of the eighteenth century and by 1900 the overall concentration had edged up to 293 ppm. The rate of increase was relatively slow until the middle of the twentieth century

Table 6.3 Atmospheric Carbon Dioxide Concentrations[1]	
	Carbon Dioxide Concentration (ppm)
1700	270–280
1900	293
1940	307
1960	312
1970	326
1980	339
1990	354
2000	369
2010	390
2050	440–500
2100	500–700
Source: US Earth System Research Laboratory.	

but by 1960 it was 312 ppm, around 32 ppm higher than 260 years earlier. By 1980 the atmospheric concentration had risen a further 27 ppm and in 2000 it was 369 ppm, an additional 30 ppm in 20 years. In 2010 the concentration was 390 ppm and in 2014 it was close to 400 ppm. The table also shows predicted ranges in 2050 and 2100 based on a variety of sources.

Additional carbon dioxide in the atmosphere causes the global average temperature to rise by restricting the amount of heat that can escape into space, a process that is commonly called the greenhouse effect. This can be magnified as global temperatures rise by factors such as the loss of ice at the poles; ice and snow reflect radiation from the sun back into space so reducing the surface area of ice at the poles, reduces this reflection.

Over the last 2000 years, the warmest period until the twentieth century was around 1000 AD but the regular fluctuations seen in global temperature over the past two millenia have been overtaken by a continuous rise since around 1900. Since then, the average temperature has risen by around 0.5–0.6°C as a result of greenhouse gas emissions according to the best modeling. Research for the UN Environment Programme[2]

[1]Data before 1959 are derived from ice core measurements. The data since 1959 are based on measurements at Manua Loa in Hawaii. Dr Pieter Tans, NOAA/ESRL and Dr Ralph Keeling, Scripps Institution of Oceanography. Predictions are based on generally proposed levels from different sources.
[2]The Emissions Gap Report 2014, The UN Environment Programme 2014.

has estimated that it is necessary to limit the global atmospheric temperature rise associated with the greenhouse gas effect to 2°C to avoid the most serious consequences of global warming. In order to achieve this, it is proposed that anthropogenic carbon dioxide emissions should become zero between 2055 and 2070.

While carbon dioxide is not the only greenhouse gas[3] it is considered to be the most important and the control of its release is considered vital. The capture and removal of carbon dioxide from fossil fuel power plant flue gases is not yet mandatory anywhere but measures to try to control its emissions are being introduced in some parts of the world. At the same time, methods for capturing the gas are being developed and there is a growing consensus that these will need to be deployed on a commercial scale after 2020 if global warming is to be limited to 2°C as noted above. If this becomes necessary then coal-fired power plants will be in the front line since they are the greatest emitters.

PRE-COMBUSTION COAL CLEANING

Most environmental processing associated with coal combustion focuses on removing pollutants after combustion has taken place. However, it is also possible to treat the fuel prior to combustion in order to reduce the quantity of pollutants it contains. As discussed earlier, there are two primary approaches to coal cleaning, physical cleaning and chemical cleaning. Chemical cleaning involves treating coal with a chemical reagent. Processes of this type are being developed but are not yet employed commercially. Physical separation relies on a difference in density between coal and it common impurities. This is applied quite widely during coal preparation. Physical cleaning can reduce the amount of sulfur in some coals and it can remove some heavy metals. However it will not eliminate the need for further processing of flue gases to these pollutants.

---

[3]Methane is considered to be 21 times more efficacious than carbon dioxide in the atmosphere and its release probably accounts for 20% of the "enhanced greenhouse effect." However, it is relatively short-lasting, remaining in the atmosphere for only 11–12 years. The concentration in the atmosphere is around 2.5 times the level before the Industrial Revolution.

COMBUSTION STRATEGIES TO REDUCE NITROGEN OXIDE PRODUCTION

Coal cleaning can help to reduce the generation of sulfur dioxide by removing the precursor, sulfur, from the coal. While nitrogen bound in coal is an important source of nitrogen oxides, managing the generation of nitrogen oxides by removing nitrogen from coal is generally not possible. Attempts have been made to chemically treat coal to remove this nitrogen but none has achieved commercial viability.

The bound nitrogen in coal comes from organic materials that originated in the plants that were coal's precursors. How much remains depends on the type of coal. Anthracite usually contains less than 1% nitrogen, bituminous coals can contain up to 3% while the amount in lignite is usually less than 2%.

The nitrogen bound in coal accounts for around 75% of the nitrogen oxides generated during combustion in a typical large pulverized coal furnace. The remaining 25% comes from the air used to supply oxygen for the combustion reaction. Since air contains around 78% nitrogen, this provides a plentiful supply.[4] One way of reducing the production of nitrogen oxides in this way is to remove nitrogen from air, so that the combustion plant is supplied with pure oxygen. This is not a strategy that has been employed simply to reduce nitrogen oxide production in pulverized coal-fired power stations, though it can be of value when considering carbon dioxide capture. However, control of the combustion process itself is used to minimize production.

Nitrogen oxides are formed when nitrogen in the coal and in air reacts with oxygen in the combustion air. The three main products are nitrogen oxide (NO), nitrogen dioxide (NO_2), and nitrous oxide (N_2O). Together, these are referred to as NO_x. Of the three individual oxides, NO is the most important, accounting for around 90–95% of the nitrogen oxide production. This is the most stable form of oxidized nitrogen at the temperatures encountered in a pulverized coal furnace. Nitrogen dioxide contributes most of the remaining 5–10%, with nitrous oxide accounting for 0.1–1% of the total.

[4]Nitrogen oxides generated from nitrogen in air are referred to as thermal NO_x while that produced from bound nitrogen is called fuel NO_x.

The formation of nitrogen oxides from both fuel nitrogen and nitrogen from air is relatively slow at low temperatures but accelerates at higher temperatures and can be rapid at the elevated temperatures in the flame of a pulverized coal burner. The reaction mechanisms involved can be complex, often involving a third component in addition to nitrogen and oxygen, a component which helps to break a chemical bond releasing reactive nitrogen atoms. However, the extent of the reaction can be limited by controlling the amount of oxygen available in the hottest part of the combustion flame. If there is too little oxygen available that which is present will preferentially react with carbon instead of nitrogen, reducing overall nitrogen oxide generation.

In a typical pulverized coal-fired boiler, coal and primary air are admitted into the furnace together through a burner. Secondary air is then added from ducts around the burner in order to manage the ratio of carbon to oxygen within the flame. If insufficient oxygen is admitted at this stage, then the combustion of the coal will not proceed to completion and it will take place under reducing conditions which limit NO_x production. Further air is then introduced around the heart of the fireball, where the temperature is lower and this allows the combustion process to finish, with the remaining carbon being converted into carbon dioxide. This type of "staged combustion" can reduce the production of nitrogen oxides by around 30–55%.

The addition of the oxygen needed to complete the combustion process can be delayed longer, allowing the combustion gases to start to cool before introducing more air higher up the furnace as the gases begin to rise and exit the combustion chamber. Using "overfire air," as this technique is called, in conjunction with more limited oxygen addition during staged combustion in the main combustion zone can reduce overall production of nitrogen oxides by up to 60%.[5] A pulverized coal boiler using these low NO_x strategies is shown in cross-section in Figure 6.1.

A third combustion strategy is called "reburning." This involves introducing further pulverized coal or even natural gas above the combustion zone, again in a region where the combustion gases have started to cool. The additional fuel will react with oxygen but under the

[5]The difference between overfire air and staged combustion is not clearly defined and they are best considered as aspects of the same process.

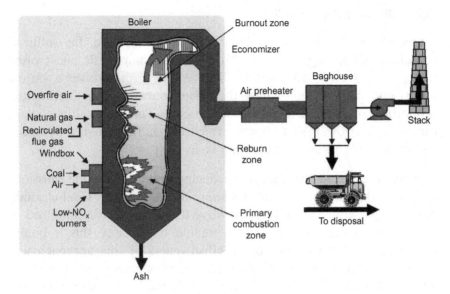

Figure 6.1 Cross-section of a pulverized coal boiler showing typical low NO$_x$ strategies. Source: Image courtesy of the US Department of Energy.

cooler, relatively reducing conditions, it will actually steal oxygen from nitrogen oxides that have already been formed. When this is combined with staged combustion and overfire air, a reduction in the production of nitrogen oxides of 70% compared to an unmodified boiler can be achieved.

Yet another means of controlling NO$_x$ formation is with flue gas recirculation. This involves taking around 20–30% of the flue gas, at a temperature of 350–400°C, and mixing it with the combustion air that is fed with coal into the burner. The effect of this is to dilute the oxygen in the combustion air, reducing the overall combustion temperature and hence reducing nitrogen oxide production.

Using these techniques will help reduce the production of nitrogen oxides but the concentration within the flue gases, even with the best low NO$_x$ combustion designs, is usually higher than emission limits prescribed for large combustion plants. The typical concentration from a large pulverized coal boiler with an optimized low NO$_x$ burner is around 700 mg/m^3. Plants with fluidized bed boilers can achieve lower levels because of the lower temperature at which combustion takes place. A large lignite-fired fluidized plant may be able to keep the NO$_x$ concentration below 300 mg/m^3. This is still well above the limit in Table 6.2, so additional measures are also required.

SULFUR DIOXIDE CAPTURE

The sulfur found in coal occurs in two forms. In one, the sulfur is contained within separate particles of sulfur compounds, most often iron pyrites, mixed with the coal. The second form involves a variety of organic carbon compounds that also contain sulfur. The actual proportion of each varies from coal to coal. The total sulfur content can be well below 1% (these are termed low-sulfur coals) but is more commonly around 2–3% and it can be as high as 10%.

There are various strategies for reducing the sulfur dioxide emissions from a coal-fired power plant. Cleaning the coal is one. Coal cleaning processes, described earlier, can remove the pyritic sulfur contained in coal but they cannot remove the organic sulfur since this is bound chemically to the carbon, so the effectiveness of this approach will depend upon the coal. Another strategy where emissions are controlled by legislation is to burn a low-sulfur coal. In the USA, for example, the production of low-sulfur coal from Wyoming has increased massively since the introduction of legislation to control sulfur emissions from coal plants. However, this can affect overall costs if low-sulfur coal has to be transported long distances.

The alternative, and the most common approach adopted today, is to capture the sulfur dioxide from the flue gases that emerge from the plant furnace. This can be carried out in a number of ways. The most common is by using a process called wet scrubbing, which involves a sorbent carried in a liquid to capture the pollutant. Where water is scarce, spray dryer absorbers using sorbent slurries which contain much less water can also be used. Both usually employ a sorbent material that is used once but there are some systems where the sorbent is regenerated and recycled through the absorption system. Dry sorbent injection systems have also been employed. Some more complex processes that involve capture of sulfur dioxide and nitrogen oxide together have been tested. Besides these, it is possible to capture sulfur dioxide directly in a fluidized bed boiler, as discussed earlier.

While a variety of chemicals will react with and capture sulfur dioxide, by far the most popular is limestone, calcium carbonate. When finely ground and carried in water, the material will react rapidly with sulfur dioxide to produce a mixture of calcium sulfite and calcium sulfate. Additional oxidation using air can convert most of the sulfite to sulfate. Limestone is extremely abundant in the earth's surface while the final product of these reactions, calcium sulfate, gypsum, can be recycled as a building material.

Limestone is most suited to wet absorption systems. For dry or semi-dry systems, calcium oxide (CaO, lime) or calcium hydroxide (Ca(OH)) are normally used. Sodium carbonate has also been employed.

In fact, the simplest and cheapest method of capturing sulfur dioxide is simply to inject a powder of finely ground lime into the hot flue gases after they have exited the power plant boiler. The particles become entrained with the flue gases, mixing with them, and the particles of calcium oxide react with the sulfur dioxide relatively rapidly at the elevated temperature. Water vapor may also be injected to aid the reaction. Provided the transit time is adequate, and the temperature high enough, the result is a flue gas laden with particles of calcium sulfite, some calcium sulfate, as well as ash from the coal, and unreacted lime. All these particles are then removed using a particle filtration system, producing a residue that is a mixture of calcium—sulfur compounds and ash. Using this method, around 30—60% of the sulfur content of the flue gases can be removed. This may be adequate if the coal already has a low sulfur content but in most cases this will not allow the flue gases to meet local emission standards. In this case a more effective method is required.

Spray drier absorption also results in a dry particular product but is more effective than simple injection. In this case the lime is first mixed with water to generate a slurry and this is then sprayed into the path of the flue gases as a cloud of fine droplets in the spray dry absorber chamber. Sulfur dioxide is absorbed in the droplets and reacts with the lime, producing calcium sulfite and calcium sulfate. Other acid gases such as sulfur trioxide and hydrogen chloride, which may also be present, will also be absorbed, while the residence time of flue gases within the absorption chamber is sufficient to evaporate the water, leaving dry particles. A schematic of a spray drier absorption system is shown in Figure 6.2.

As with the sorbent injection system, the particles must be removed from the flue gas using a filtration system. If this is fitted after the spray drier absorber, then the material collected will be a mixture of fly ash and

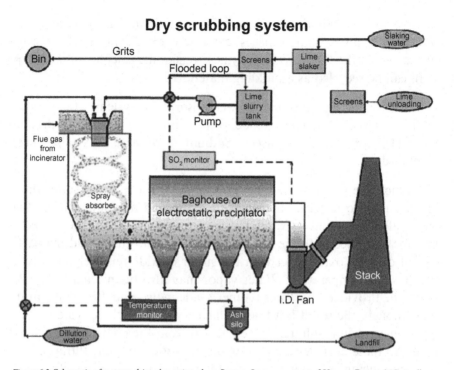

Figure 6.2 Schematic of a spray drier absorption plant. Source: Image courtesy of Hamon Research-Cottrell.

calcium/sulfur compounds as with the dry injection system. An alternative, commonly employed in Europe when this system is used, is to have two particle collection systems. The first, before the spray drier absorber, collects all the fly ash while the second collects the calcium sulfur compounds formed by sulfur capture. This results in a much purer product that can be more readily recycled for building or other purposes.

The spray drier absorber system is simpler and cheaper than the main alternative high-performance sulfur removal system, wet scrubbing, but it is less effective and the reagents are more expensive. It is often used on smaller power plants but less so on large facilities and accounts for 12% or less of all the coal plant capacity fitted with sulfur capture.

The most effective, and the most popular, system for capturing sulfur dioxide is wet scrubbing. This is similar to the dry absorption techniques outlined above but employs a liquid rather than a slurry or dry powder as the capture reagent. The capture reagent in wet scrubbing is usually powdered limestone although calcium hydroxide can also be used.

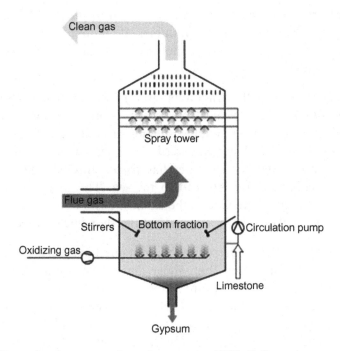

Figure 6.3 Cross-section of a spray tower. Source: Image courtesy of Wikipedia Commons.

In a typical wet flue gas desulfurization (FGD) system limestone is ground to a fine powder and them mixed with water. Typically this "lime wash" contains 10% limestone. The lime wash is pumped through nozzles and sprayed down from the top of a spray tower through which the flue gases travel from the bottom to the top. The hot flue gases evaporate some of the water which exits with the flue gases as water vapor. Meanwhile the sulfur dioxide (and some other acidic gases) dissolve in the water and then react with the limestone particles to produce calcium sulfite. The liquid is collected at the bottom of the tower and pumped away. In many wet FGD systems air is blown through the resulting calcium sulfite solution to promote its oxidation to calcium sulfate, a process called forced oxidation. This provides a more saleable product. The resulting gypsum is then separated from water and can be sold. Figure 6.3 shows a cross-section of a typical spray tower.

A wet scrubbing system can capture between 90 and 98% of the sulfur in the flue gases from a coal plant. With additives that can be pushed as high as 99%. The process is complex, akin to adding a chemical processing plant to a coal-fired power plant, but is currently the most effective method available. It is used on many large coal-fired power plants across the globe and usually forms part of every new coal station.

An alternative to the traditional wet FGD scrubbing system is a seawater FGD system. Seawater is naturally alkaline and will absorb sulfur dioxide, producing soluble sulfates and sulfites. The process is similar to the wet FGD system described above, with seawater replacing the lime wash. After the seawater has passed through the absorption tower it is collected and returned to the sea. This system can remove 98−99% of the sulfur dioxide in the exhaust gases of a coal plant but is only feasible at plants that are located at the shoreline of a sea.

A range of other, more complex, systems have been tested in the past for sulfur removal but most have not found commercial application. One of the more interesting of these is the use of activated carbon or charcoal. The can be made from coke and is capable of absorbing pollutant gases such as sulfur dioxide and nitrogen oxide onto its surface. It can remove other pollutants too, such as hydrogen chloride and mercury. The spent activated carbon is then regenerated with the release of the absorbed sulfur which can be extracted as pure sulfur.

NITROGEN OXIDE CAPTURE

While low NO_x combustion strategies discussed earlier can help reduce the quantity of nitrogen oxides found in the flue gases from a coal-fired power plant, they cannot reduce it to a level that will comply with emission limits in most jurisdictions. As a consequence, further action is necessary to reduce the NO_x concentration below the legal limit. Two processes are commonly used for this purpose. Both involve adding a reagent, usually either ammonia or urea, to the flue gases then encouraging it to react with the nitrogen oxides to reduce them back to molecular nitrogen. One relies on the temperature of the flue gases to achieve its aim, while the other uses a metallic catalyst to promote the reaction between the reagent and the NO_x.

The first, and simplest, of these methods involves the injection of ammonia or urea into the flue gases soon after they leave the boiler, when the temperature is between 870°C and 1200°C.[6] Between these temperatures the ammonia will react spontaneously with nitrogen oxides to produce a mixture of nitrogen and water vapor which is then carried away with the flue gases. This process is generally called selective non-catalytic reduction (SNCR). It can remove between 35 and 60% of the nitrogen oxides produced during combustion.

SNCR will be effective if the nitrogen oxide concentrations are already low. However, it can lead to ammonia contamination of fly ash and to the release of ammonia into the atmosphere with the flue gases, a phenomenon called ammonia slip. This can be controlled using feedback to ensure the amount of ammonia being injected is just sufficient for the concentration of NO_x present but ammonia slip is always a problem. The technique has been in use since the mid-1970s in Japan and it is also used in Europe and the USA. It is best suited to smaller plants as it can be difficult to implement effectively in very large coal-fired boilers.

The alternative process, which is called selective catalytic reduction (SCR), requires ammonia or urea to be mixed with the flue gases and then passed over a solid catalyst that facilitates the reaction, converting NO_x into molecular nitrogen and water vapor. The process normally takes place at a point where the flue gases are between 300°C and 400°C. The catalytic process is very efficient and allows precise control of the amount of ammonia to match the NO_x present, minimizing the amount of ammonia slip. An emission reduction of 80–90% can be achieved.

There are three main configurations for SCR. The first, called high dust, involves placing the SCR reactor before the dust collection system in the plant. Figure 6.4 shows the typical layout. This is the most common arrangement, particularly with dry bottom boilers where little slag is generated. However, the fly ash can erode the solid catalyst in the SCR reactor so in some plants a low dust configuration is adopted with the SCR reactor placed after the dust collection system. The disadvantage of this is that the dust collection must occur at a relatively high flue gas temperature in order to ensure the gases reach the SCR reactor at a high enough temperature. Another low dust configuration, called tail-end SCR, provides a more compact design.

---

[6]If the temperature is too high, ammonia starts to decompose thermally, if it is too low the reaction rate is too slow.

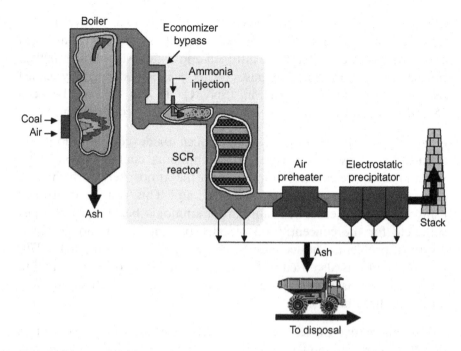

Figure 6.4 Schematic of a selective catalytic reduction system in a coal-fired power plant. Source: Image courtesy of the US Department of Energy.

The basic form of the catalyst for SCR is most commonly a titanium oxide into which vanadium has been dispersed as the active catalytic ingredient. However, most catalysts are proprietary and they can have much more complex structures with other elements added to improve efficacy. Activate carbon can also be used as the catalyst, often carried on a moving bed.

One of the problems that can be caused by SCR systems is the production of sulfur trioxide from any sulfur dioxide present in the flue gases. This extremely acid gas forms sulfuric acid in contact with water, making it highly corrosive. As a consequence of this SCR was initially used only for plants burning low-sulfur coals. However, with effective wet scrubbing FGD systems it has been possible to apply SCR to plants that burn high-sulfur coals.

In addition to these two methods, it is possible to remove NO_x using a wet scrubbing system similar in concept to that described for wet FGD systems above. The reagent sprayed into the flue gas stream in this case is a solution sodium hydroxide, although hydrogen peroxide has also been used. The reaction that takes place leads to the formation of soluble nitrites and nitrates which must be isolated for disposal. Wet scrubbing is not normally used in power stations for NO_x removal.

COMBINED SULFUR AND NITROGEN OXIDE REMOVAL

The capture of sulfur dioxide and that of nitrogen oxides, each involve addition of a complex chemical system to a coal-fired power plant. If these could be combined into a single process it would be a major simplification and could reduce costs significantly. Several attempts have been made to develop combined systems of this type but none has achieved major commercial success.

One, already mentioned, involves using an activated carbon as the capture agent. Carbon particles, usually formed from coke, are extremely porous and provide a very large surface area, while the carbon itself will readily form bonds with many compounds. If the activated material is slowly cycled through a tower, up which the flue gases pass, nitrogen oxides, sulfur dioxide, metals such a mercury, and other acidic gases such as hydrogen chloride, will be adsorbed onto the surface of the carbon.

The bond formed between the carbon and these molecules is not strong and if the activated adsorbent is removed from the tower and then heated in a separate reactor, the pollutant molecules will be released and can be collected and processed while the carbon can be recycled. In one version of this system, sulfur can be recovered as a by-product. However, the system has never been able to prove itself commercially.

Another approach that has been tried is to use a beam of electrons to convert and capture pollutant molecules. In this system, the flue gases are exposed to an intense beam of electrons in the presence of ammonia gas. The electrons activate both sulfur dioxide and nitrogen oxides, allowing them to react with ammonia to produce a mixture of ammonium sulfate and ammonium nitrate, which can be sold as a fertilizer. This system was initially developed in Japan and later tested in Germany but has never been marketed commercially.

More recently, scientists have explored the possibility of oxidizing nitrogen oxides in the flue gases using either ozone or hydrogen peroxide injection, leaving a range of higher oxides of nitrogen which will then react with limestone or lime in a wet FGD scrubber of spray drier absorber. This process, if developed commercially, is likely to be targeted at small boilers and waste incineration plants where it can be used to capture a range of pollutants at relatively low cost.

PARTICULATE (DUST) REMOVAL

In most large coal-fired power stations a large part of the ash residue from coal combustion is carried away with the flue gases, entrained in the form of fine particles. If allowed to escape into the atmosphere these will create a plume of smoke before eventually falling to earth as a layer of fine dust. Modern emission regulations require that this material be captured before flue gases are allowed to exit the stack of the power plant.

There are two principal systems that are used for removing particulates from the flue gas of a coal-fired power station, electrostatic precipitators (ESPs) and fabric (baghouse) filters. Cyclones can also be used to capture particulate material but they tend to be used on smaller plants.

The ESP was invented by the American scientist Frederick Cottrell. It utilizes a system of plates and wires to apply a large voltage across the flue gas as it passes through the precipitator chamber. The flue gases, when they enter the ESP, pass through an array of wires that are held at a high negative voltage relative to ground so that they generate a corona of ionized gas around them. As the particles pass through this corona they become charged. Once charged they are attracted to large vertical plates that are held at ground voltage. This is shown in Figure 6.5. Periodically the plates are vibrated or "rapped" causing the layer of charged particles that has built up on the surface of each to fall into a collector at the bottom of the ESP.

ESPs are extremely efficient. A new ESP will remove between 99.0% and 99.7% of the particulates from flue gas. However, it must be tuned to the particular coal being burned in the power plant. Where coals of different types and from various sources are to be burnt, the alternative may be more effective. The ESP can handle both dry and wet particles but it is less effective when the ash has a high electrical resistivity.

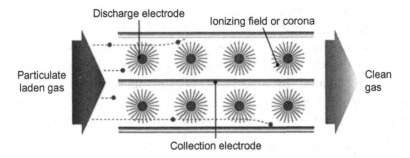

Figure 6.5 Operating principle of an electrostatic precipitator. Source: Image courtesy of Powerspan Corp.

Bag filters, or baghouses, are tube-shaped filter bags through which the flue gas passes on its way to the power plant stack. Particles in the gas stream are trapped in the fabric of the bags from which they are removed using one of a variety of bag-cleaning procedures. These include using supersonic blasts of air to dislodge particles so that they fall to the base of the unit and can be removed. These filters can be extremely effective, removing over 99% of particulate material. They are generally less cost-effective than ESPs for collection efficiencies up to 99.5%. Above this, they are more cost-effective. A system that combines a baghouse-style filtration system with an ESP is under development too. This aims to provide a cost-effective high removal efficiency system, but has not yet been extensively demonstrated.

Baghouse filters require regular replacement, which can make them expensive to operate. ESPs are expensive to construct but are much more economical to operate.

Cyclone filters, which capture particles by imparting a centrifugal force on them, are used in CFB boilers to trap particles escaping the combustion zone but they are not usually used to capture particles in large power plants. They are relatively cheap to build, are able to operate at high temperatures and have low maintenance costs. However, they are only effective with dry particles and they are not efficient at removing small particles from flue gases.

MERCURY REMOVAL

Most coals contain a small amount of mercury and this can easily end up being discharged in the flue gas from a coal-fired power plant.

The metal has an impact on many parts of the human body and exposure is always harmful. According to the US Environmental Protection Agency coal-burning power plants are the major anthropogenic source of mercury in the environment. The organization has estimated that 50% of the mercury released in US coal plants falls on the USA with the rest being carried further afield. Globally, mercury from coal combustion plants is likely to be spread across all continents.

The regulation of mercury emissions from coal plants has been under discussion for many years. Regulations controlling the level of emissions from power plants are due to be introduced this decade and may necessitate additional measures to ensure that limits are not exceeded.

Dust removal systems in power plants will generally remove around 25% of the mercury released during combustion. When a wet scrubbing sulfur removal system is also installed this can increase to 40%–60%. Adding SCR can lead to 95% removal with bituminous coals. However sub-bituminous coals and lignites do not respond so well so alternative measures may be needed to reduce mercury emissions to below regulatory limits.

The injection of activated carbon particles has been used to remove impurities, such as mercury, in waste incineration plants and this appears to offer the best solution where further mercury capture is necessary. The carbon particles will then be removed in the dust removal system through which the flue gases pass at a later stage. It is expected that plants will eventually need to remove 90% or more of the mercury released during combustion.

CHAPTER 7

Carbon Capture and Storage

There are currently no commercial carbon capture technologies that can be applied to large coal-fired power stations, although several configurations are being tested and most of the technology required is well established. However, legislation and incentives to limit carbon emissions are expanding with systems such as the carbon certificate policy and emissions trading system within the European Union where large emitters are required to have carbon certificates for every tonne of carbon they emit.

Faced with growing pressure to control their emissions, utilities and independent power generators are attempting to find strategies other than carbon capture that will reduce the amount of carbon dioxide they produce. Increasing plant efficiency is one strategy and there has been a slow global shift in the last decade or two towards supercritical and ultra-supercritical boilers in coal-fired plants. These produce less carbon dioxide per unit of electricity than a conventional sub-critical boiler.

A more common strategy, particularly in the developed world, has been to shift away from coal altogether and generate power from natural gas. The latter produces significantly less carbon dioxide for each unit of power than the best supercritical coal-fired plant. However, gas is both more expensive and exhibits greater price volatility than coal, making the economic risk associated with investing too heavily in gas-fired plants very high. That, coupled with the easy availability of coal, particularly in countries like China and India, means that coal plants will remain crucial to global power generation at least until the middle of this century, if not longer. Faced with this, the development of new strategies to reduce carbon emissions from coal plants is vital if atmospheric warming is to be limited. That means carbon capture and sequestration.

For carbon capture there are three main approaches available. The first, often called post-combustion capture, involves installing a

Coal-Fired Generation. DOI: http://dx.doi.org/10.1016/B978-0-12-804006-5.00013-7

plant similar to an FGD scrubber to the exhaust of the power plant. Reagents capable of capturing carbon dioxide when deployed from a spray tower are already available and this technology is likely to be one of the simplest and possibly the most economical to deploy. An alternative to this, called pre-combustion capture, involves pre-treating coal to remove the carbon before combustion. This is achieved using a modified version of coal gasification which leaves a fuel gas composed primarily of hydrogen. The gasification process also requires some form of carbon dioxide capture technology and this is likely to be via a scrubber system too, although in this case other solutions may be possible. The third way of tackling carbon dioxide emissions is to sidestep the difficulty of separating carbon dioxide from flue gases, which are usually a mixture of residual oxygen, nitrogen, carbon dioxide, and other trace gases. Instead, oxygen is separated from air first, and then coal is burnt in virtually pure oxygen. The scheme, called oxyfuel combustion, leads to an exhaust gas stream composed primarily of carbon dioxide, mixed with some water vapor and excess oxygen from which it is much easier to isolate the carbon dioxide than when the latter is mixed with nitrogen. In effect, the oxyfuel system replaces carbon dioxide separation with oxygen separation.

If carbon dioxide is captured from coal before or after power generation, the gas must then be stored in a way that prevents it ever returning to the atmosphere. This is called carbon sequestration. The most likely place for carbon sequestration is underground and some pilot scale projects have demonstrated that this is feasible. However, the process will have to be carried out at an extremely large scale so this technology needs to be perfected alongside carbon capture technologies if carbon capture and storage is to become feasible.

There is one last strategy that can be used to reduce the environmental impact of coal combustion and that is to replace some of the coal burnt in the coal plant with a biomass fuel such as wood chips. The biomass, provided it comes from a sustainable source, is essentially carbon neutral so while the combustion of the biomass still releases carbon dioxide, this should be recaptured from the atmosphere when further biomass is grown. The process, called biomass cofiring, cannot eliminate carbon emissions from coal plants and so does not offer a long-term solution — unless the coal plant converts to 100% biomass firing.

BIOMASS COFIRING

The cofiring of biomass in a coal-fired power plant involves replacing some of the coal with a biomass fuel. The fuel can be burned in any type of coal boiler including pulverized coal-fired power stations and fluidized plants. However, all plants will require some adaptation to make cofiring possible.

Cofiring will reduce the environmental impact of a coal plant by reducing the net amount of carbon dioxide added to the atmosphere. The biomass will generate carbon dioxide when burnt in the same way as coal but subsequent re-growth of biofuel will absorb carbon dioxide from the atmosphere again so that the net contribution is zero. This relies on supplying fuel from a sustainable and continuous biomass production source. Some fuel sources, particularly where the fuel is transported over long distances, provide a questionable environmental advantage. Nevertheless, if managed properly it does provide a potential environmental gain.

One of the main advantages of cofiring is that it allows biomass to be burnt in a power plant that operates at high efficiency. Most dedicated biomass combustion plants are small and have relatively low energy conversion efficiencies. Modern, large coal-fired plants on the other hand offer some of the best combustion efficiencies achievable. There is very little loss in efficiency when part of the coal is replaced with biomass, so burning biomass in this way allows significant gains compared to the use of more conventional biomass combustion plants.

Small amounts of biomass can be mixed with pulverized coal and introduced into the boiler furnace through the coal burners. This appears to be effective for up to around 10% cofiring, although it is normally only used for up to 5% cofiring. Up to this level, the biomass fuel can simply be mixed with the coal before it enters the power plant processing train allowing the two to be processed together in the coal mills and then fed to the plant burners.

When larger proportions of cofiring are planned a separate biomass preparation system is needed to pre-mill the biomass before it is mixed with the milled coal and injected into the boiler. With this type of system it appears possible to use up to 20% biomass cofiring without major modification to the plant burners and this is the scheme adopted in most coal-fired plants in Europe that have experimented with large-scale biomass cofiring. The alternative, when large-scale cofiring is being considered, is to fit separate burners to introduce the biomass fuel into the furnace.

For high cofiring ratios it may be possible to modify the biomass fuel by a process called torrefaction. This involved heating the fuel to around 200–300°C in a reducing atmosphere for around 1 h. The torrefaction alters the properties of the biomass, making it more like coal in terms of handling and preparation. The technique is still in the demonstration stage. Its aim is to allow biomass to be burnt in a coal-fired power plant with only minimal modification. This might allow 100% biomass firing without major plant modification.

There are alternative approaches to cofiring. One is to gasify the biomass fuel first and then burn the gaseous product in a boiler. However, this appears to be more expensive and may not be an economically viable approach, while adding significant complexity compared to the simple cofiring outlined above.

Fuels for cofiring include woods and grasses, both of which can be grown as dedicated biomass fuels. One potential problem with many biomass fuels is that they can cause fouling of boilers that have been designed to burn particular coals. Woods often have similar low-ash, low-alkali, and low-chlorine content as coal fuels and these should not present an increased problem. However, herbaceous fuels, such as grasses, may have higher levels of ash, alkali, and chlorine. These can lead to higher levels of corrosion within boilers.

One potential advantage of cofiring is that it can lead to lower nitrogen oxide production because the biomass fuel contains less fuel nitrogen than the coal it is replacing. While some gains have been seen from biomass cofiring, the relationship between the amount of nitrogen in the fuel and NO_x production is not simple. In addition, high levels of alkali metals in biomass can lead to early poisoning of the catalyst in an SCR reactor, leading to the need for early replacement. Sulfur dioxide emissions may also be lowered as biomass fuels contain only low levels of sulfur. However, the effects on particulate emissions can be more complex.

There is one further interest in co-firing. When the technology is combined with carbon capture it can lead to a net reduction in the amount of carbon dioxide in the atmosphere. This is because carbon dioxide is being removed from the carbon neutral part of the combustion fuel, leading to a (long-term) net fall in the amount of carbon dioxide. If widely adopted in the future, this could offer a means of actually reversing the build up of carbon dioxide within the atmosphere.

POST-COMBUSTION CAPTURE OF CARBON DIOXIDE

The post-combustion capture of carbon dioxide in a coal-fired power plant is conceptually the simplest scheme for controlling carbon emissions. The scheme, which involves adding a plant that will capture and remove carbon dioxide from the flue gases that emerge from the plant boiler, before they are released into the atmosphere, is also the easiest type of carbon capture plant to fit to existing coal-fired power plants. This is likely to be an important consideration if carbon capture from coal plants becomes mandatory.

The capture process is conceptually very similar to that used in FGD scrubbers to remove sulfur dioxide. A reagent capable of capturing carbon dioxide is sprayed from the top of a tall reactor vessel, the absorption tower, through which the flue gases pass from bottom to top. As the reagent cascades down the tower it mixes with the plant flue gases and carbon dioxide is removed by being bound chemically by the reagent. This type of process is capable of removing up to 90% of the carbon dioxide contained in the plant exhaust gases.

The spent reagent is collected at the bottom of the absorption tower while the cleaned flue gases are allowed to escape to the atmosphere. The reagent must now be treated to release the carbon dioxide, leaving the capture agent to be recycled through the system. This is an energy-intensive process because in this type of system the carbon dioxide molecules are relatively strongly bound to the reagent and it can lead to an overall fall in efficiency, using the materials available today, of close to 28%. This means that a supercritical coal-fired plant with an efficiency of 45% would see its overall efficiency drop to 33% or lower once carbon capture had been added. Once the carbon dioxide has been released, it must then be compressed ready for transportation, expected to be by pipeline, to a facility where it can be sequestered. A schematic of a power plant with post-combustion capture is shown in Figure 7.1.

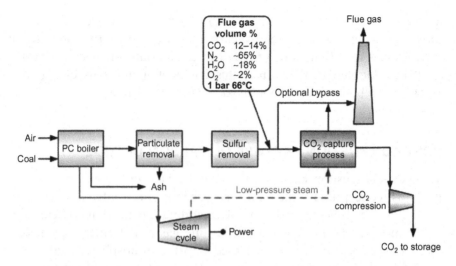

Figure 7.1 Schematic of a power plant with post-combustion carbon dioxide capture. Source: Image courtesy of US Department of Energy.

The efficiency of carbon dioxide capture depends on the concentration of carbon dioxide within the flue gases of a power plant. For a typical coal-fired boiler the concentration is normally between 12% and 14% of the flue gas by volume. At this concentration the most effective way of capturing carbon dioxide is to use a reagent that bonds chemically with the carbon-containing gas. The chemical solvent will remove a large part of the carbon dioxide but, because it binds strongly with it, releasing the gas again is more difficult. In other situations where the concentration of carbon dioxide is much higher it is possible to employ a physical solvent into which the carbon dioxide simply dissolves. The bonds holding the gas in solution are much weaker than in the case of the chemical solvent and the energy required to release it again is much smaller.

The best solvents for capturing carbon dioxide chemically are amines, which bond with it strongly. Amines have been in use since the 1930s to produce food-grade carbon dioxide from gas streams with between 3% and 25% carbon dioxide. Of these, the most commonly used is monoethanolamine (MEA). The technology for MEA capture is well tested but it has not yet been tried at the level of a large coal-fired power station. An alternative to MEA is an ammonia-based solvent from which it is potentially easier to release carbon dioxide once it has been captured. However, capture efficiency will be the key to its success.

Other types of capture have been proposed for post-combustion capture. These include the use of ionic liquids which can, in principle, absorb carbon dioxide without binding to it very strongly. Cryogenic separation at a temperature of $-120°C$ or lower may be capable of removing 90% of the carbon dioxide from the flue gas stream but cooling costs are the key consideration. Solid sorbents which adsorb carbon dioxide onto their surface are attractive if they can provide a high enough capture efficiency. Today they cannot. Membranes, similarly, would offer a simple solution if they could achieve the required efficiency.

PRE-COMBUSTION CAPTURE OF CARBON DIOXIDE

The pre-combustion capture of carbon dioxide is essentially coal gasification as described earlier, with the gasification processes carried through so that the product gas is essentially carbon dioxide and hydrogen. To carry out gasification, coal is first heated in a special reactor in a mixture of either air or water and steam normally under pressure. Most will use oxygen so that the plant will need an air separation unit (ASU) to supply oxygen. The product of this reaction will be a mixture of compounds including hydrogen, carbon monoxide, carbon dioxide, and methane. The gas mixture, called synthesis gas, is then mixed with more steam and passed over a catalyst which promotes the reaction between carbon monoxide and water, the water shift reaction, producing carbon dioxide and hydrogen.

Various gasifiers have been developed for coal gasification including fixed bed gasifiers, fluidized bed gasifiers and entrained flow gasifiers. The last two are the most suitable for power plant use. Both are operated at around 25–30 bar. An entrained flow gasifier operates at a much higher temperature than a fluidized bed gasifier (it is more like a pulverized coal furnace) and requires more oxygen but also produces a higher-quality gas product. After gasification the product gases must be cleaned to remove dust, acid components such as hydrogen chloride and hydrogen sulfide,[1] and any nitrogen oxide and carbon dioxide. This leaves a stream of hydrogen which can then be used in a power plant to generate electricity.

[1]Under the reducing conditions in a gasification reactor sulfur is converted into hydrogen sulfide rather than sulfur dioxide.

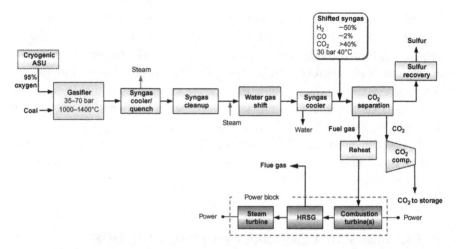

Figure 7.2 Schematic of a power plant with pre-combustion capture of carbon dioxide. Source: Image courtesy of the US Department of Energy.

The most commonly proposed configuration for a coal gasification power plant is the IGCC power plant. In this type of plant the hydrogen generated from gasification is burnt in a gas turbine to generate electricity. Waste heat from the exhaust of the turbine is then used to generate steam in a heat recovery steam generator and this drives a steam turbine. The gasification and power generation elements of such a plant are tightly integrated so that waste heat from the gasification process is also utilized for steam generation. A schematic for a plant of this type is shown in Figure 7.2. Research suggests that it may be possible to achieve up to 55% energy conversion efficiency with an IGCC plant before carbon capture but the best efficiency demonstrated by IGCC plants in operation is closer to 40%. When carbon capture is added, the overall efficiency drops to 32%. It may be more efficient to utilize the hydrogen in a fuel cell instead.

The key to the pre-combustion plant, aside from the gasification process, is carbon capture. The product gas after the water shift reactor will often contain more than 50% carbon dioxide. This high concentration makes the gas easier to separate from the hydrogen. In addition the gas is already at high pressure, and there is little nitrogen present so the volumes of gas that have to be handled are smaller. This makes is possible to carry out separation using a physical absorption technique that is much less energy-intensive than the post-combustion amine capture technique.

Physical separation of carbon dioxide is common in both the natural gas and petrol refining industries and two commercial processes are in use. One is called the Rectisol process and uses refrigerated methanol to absorb carbon dioxide. The second, called Selexol, uses a glycol-based solvent manufactured by Dow Chemicals. Both processes can also remove hydrogen sulfide, which can be isolated separately after absorption, so that finally a stream of pure carbon dioxide is produced ready for compression and transportation. The overall energy loss for a Selexol-based system is around 20%.[2] While these processes are well established, new solvents are being sought that might offer lower energy loss. Research is looking at ionic fluids and ammonia-based solvents as possible candidates.

Solvent capture requires the product gases be cooled before treatment. High-temperature absorption would avoid this energy loss and this is another avenue for development. Solid adsorbents offer one possible high-temperature solution as do membranes, but research into these is at an early stage.

CARBON CAPTURE AND OXYFUEL COMBUSTION

Oxyfuel combustion, in which combustion air in a power plant is replaced with pure oxygen, is the third main approach to coal combustion and carbon capture being developed for coal plants. It offers a radically different approach to the problem. A conventional coal-fired power plant with post-combustion capture has to separate carbon dioxide from a flue gas mixture in which there is a large amount of nitrogen. The oxyfuel process, by removing the nitrogen at the start, so that coal burns in oxygen instead of air, sidesteps this problem because the flue gases from the combustion process are largely carbon dioxide. This makes is much easier to isolate.

This simplification is at the expense of an ASU, which the plant requires to provide oxygen. In addition, the exhaust gases from the plant are not pure. They will contain some residual nitrogen, unburnt oxygen, sulfur dioxide, nitrogen oxides, and particulate material. Much of this must still be removed to produce carbon dioxide that is pure enough for sequestration.

[2]*DOE/NETL Advanced Carbon Dioxide R&D Program: Technology Update*, September 2010, NETL/DOE.

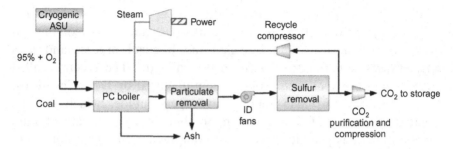

Figure 7.3 Schematic of a power plant using oxyfuel combustion for carbon dioxide capture. Source: Image courtesy of the US Department of Energy.

There is also a combustion problem. When coal burns in pure oxygen the flame temperature is much higher than when it burns in air. The peak temperature can reach 2500°C compared to 1700°C in a supercritical boiler. There are no boiler and burner construction materials today that can withstand such a high temperature. The solution to this problem is to take some of the carbon-dioxide-rich flue gas from the exit of the boiler and mix it with the oxygen fuel supplied to the burners. This dilutes the oxygen and reduces the flame temperature to a level similar to that found in a conventional air-blown plant. A schematic of an oxyfuel plant is shown in Figure 7.3.

One of the keys to the economics of an oxyfuel combustion plant is the ASU. Air separation is a well-developed technology and is normally based on cryogenic separation of the components of air but it is expensive. The overall efficiency of a supercritical boiler with oxyfuel combustion is expected to be around 26%. New air separation technologies might be able to reduce this cost but all are at an early stage of research.

Oxyfuel combustion could potentially be retrofitted to existing coal boilers. It is only likely to be economically effective on modern supercritical and ultra-supercrital plants and some adaptation is necessary to allow flue gases to be fed back to the oxygen feed system. However, it may prove an economical competitor to the retrofitting of post-combustion capture in the future.

CHEMICAL LOOPING

Chemical looping is an advanced form of oxyfuel combustion that does away with the need for an expensive ASU. In its place it uses a

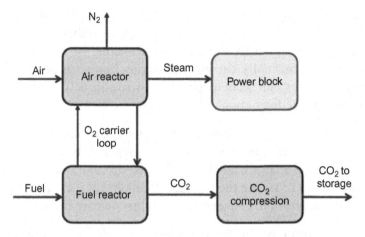

Figure 7.4 Chemical looping. Source: Image courtesy of the US Department of Energy.

solid-state oxygen carrier. The process requires two reactors. In the first the oxygen carrier is heated in the presence of air, when it preferentially captures oxygen, leaving a stream of nitrogen. The oxygen-rich solid carrier is then transferred to a second reactor where it is exposed to coal that has been gasified to produce syngas. The oxygen from the carrier will react with the combustible components of the syngas producing hot exhaust gases that contain a high concentration of carbon dioxide. This must then be purified before compression and transportation. A simple schematic of the chemical looping process is shown in Figure 7.4.

A number of materials have been identified that could potentially act as oxygen carriers in a chemical looping process. These include copper oxide, cobalt oxide, and di-manganese trioxide. Each of these oxides will capture oxygen at high temperature ($800-1200°C$) to produce a higher oxide of the metal CuO/Cu_2O_3, Mn_2O_3/Mn_3O_4, and CoO/Co_3O_4. The process is still in an early stage of development.

CARBON DIOXIDE COMPRESSION AND TRANSPORTATION

Once carbon dioxide has been captured using one of the cycles outlined above it has to be transported to a site where it can be stored securely. Transportation will probably be by pipeline, though in some cases it may involve bulk freight carriers by rail or sea. Whichever is used, the carbon dioxide must first be compressed. Compression is carried out in stages so that eventually the carbon dioxide reaches its critical point where its volume decreases significantly compared to the normal gaseous phase. Typical compression is to 10–15 MPa. Dehydration is normally carried out during compression because the presence of water can lead to pipeline corrosion as well as problems with solid hydrate formation.

Gas compression is a well-tested technology but it has an energy cost. Estimates suggest that the electrical energy required to compress the carbon dioxide from a large coal plant would use around 7–12% of the plant output. In order to reduce the cost new technologies such as refrigeration and pump compression, or shockwave compression are being explored.

Transportation of the compressed gas will most commonly be via pipeline. Ideally the distance the carbon dioxide has to be transported to a sequestration site is as short as possible but even so it is likely to be tens or hundreds of kilometers in many cases. Transportation by pipeline is carried out regularly in the oil industry, particularly in the USA where 40 m tons are pumped annually through 6000 km of pipeline.

CARBON DIOXIDE SEQUESTRATION

The capture and storage of carbon dioxide from power stations is going to involve massive quantities of the gas. An estimate based on one of the International Energy Agency future energy consumption scenarios puts the quantity that will need to be captured and stored from power plants by 2050 at around 80 Gtons.[3] There are only two places that such large amounts could feasibly be stored, firstly within or below the world's oceans, or secondly under the ground. Ocean storage is currently considered to carry too high a risk to be practical so storage will take place underground.

Three principal types of underground storage site have been identified. The easiest to access immediately are oil and gas wells. Carbon dioxide injection is already carried out by the hydrocarbon industry for enhanced oil recovery (EOR) from oil and gas fields. Evidence from such sites suggests that the gas remains underground once it has been pumped there. Continued EOR will consume a small

[3]*Technology Roadmaps: Carbon Capture and Storage*, IEA, 2010.

amount of captured carbon dioxide. More importantly, spent oil and gas fields in many parts of the world will offer a simple solution for initial carbon dioxide sequestration projects by providing a storage capacity that will often have pipeline access. The capacity that these can accommodate is limited, however, and they cannot be found everywhere, so other sites will also be required when carbon capture and storage starts to be deployed widely.

Another option is to store carbon dioxide in coal-beds that are too deep to be mined. These deep coal-beds often contain methane. If carbon dioxide is pumped down into a coal-bed containing methane it will selectively displace the gas, which can be recovered for use as an energy source above ground. The recovery of methane will offset the cost of carbon dioxide sequestration but sites of this type are relatively rare so they cannot provide for all the carbon dioxide that needs to be sequestered either.

Over the longer term the main solution to the problem of carbon dioxide sequestration is likely to be the third, and potentially the most important type of underground site, called a brine aquifer. A brine aquifer is created when a cap of impermeable rock is formed deep underground. This cap prevents water or gas seeping upwards through it and so a hot, concentrated brine collects within the more porous rock beneath the cap. This is the brine aquifer. The brine can be displaced by carbon dioxide if the latter is pumped down into the aquifer. Experimental sites where carbon dioxide has been stored in brine aquifers indicate that the gas should remain securely stored away and it may eventually be trapped permanently by a reaction with the rock in which it is held to make carbonates.

It should be noted, when considering the large-scale capture and storage of carbon dioxide, that the gas can be highly toxic. The normal concentration in the atmosphere is 0.04%. If the concentration rises to 3%, inhalation begins to effect vision and hearing. At 10%, the gas acts as an asphyxiant and when the concentration rises to 20% inhalation leads to rapid death. It is vital, therefore, that large quantities of the gas, once stored, can never be released into the atmosphere.

For this reason, as well as to verify that storage is permanent and that carbon dioxide that has been stored is not returning to the atmosphere, storage sites will need to be monitored for tens, if not hundreds of years, to verify that the storage is secure. Regulations to ensure this happens are already being mapped out. Operational sites will most likely be inspected regularly but remote monitoring is likely to become normal once a site has been sealed.

Other types of sequestration are being explored too. One of the potentially most interesting is biological sequestration. Plants capture carbon dioxide from the atmosphere during photosynthesis and this process might be harnessed to absorb captured carbon dioxide too. Research has indicated that algae might be able to process carbon dioxide rapidly, but development is at an early stage. Another option is to try to convert carbon dioxide directly into a solid mineral. Magnesium hydroxide solution will absorb carbon dioxide, producing magnesium bicarbonate. However, like biological sequestration, this remains at a very early stage.

A number of projects around the world have demonstrated that underground storage of carbon dioxide is feasible. Several of these involve brine aquifers, while others are based in EOR projects where the fate of the underground carbon dioxide has been monitored. However, as will all current aspects of carbon capture and storage, this has to be demonstrated successfully at the scale of a large coal-fired power station before the technology can be considered mature.

CHAPTER 8

The Cost of Electricity Generation from Coal-Fired Power Stations

The cost of electricity from a power station depends on a range of factors. First there is the cost of building the power station and buying all the components needed in its construction. In addition, most large power projects today are financed using loans so there will also be a cost associated with paying back the loan, with interest. Then there is the cost of operating and maintaining the plant over its lifetime. In the case of a coal-fired power station, or indeed any power station that relies on a combustion fuel, there is the cost of buying and transporting the fuel to the plant. There are costs associated with the management of any waste materials from the plant, which in the case of a coal-fired power station will include some boiler slag, waste from the desulfurization unit, ash from the particulate collection system, and − in the future − carbon dioxide. Finally, the overall cost equation should include the cost of decommissioning the power station once it is removed from service.

It would be possible to add up all these cost elements to provide a total cost of building and running the power station over its lifetime, including the cost of decommissioning, and then dividing this total by the total number of units of electricity that the power station produced over its lifetime. The result would be the real lifetime cost of electricity from the plant. Unfortunately, such a calculation could only be completed once the power station was no longer in service. From a practical point of view, this would not be of much use. The point in time at which the cost-of-electricity calculation of this type is most needed is before the power station is built. This is when a decision is made to build a particular type of power plant, based normally on the technology that will offer the lowest-cost electricity over its lifetime.

In order to get around this problem economists have devised a model that provides an estimate of the lifetime cost of electricity before the station is built. Of course, since the plant does not yet exist, the

Coal-Fired Generation. DOI: http://dx.doi.org/10.1016/B978-0-12-804006-5.00014-9

model requires a large number of assumptions and estimates to be made, from the cost of construction to the future cost of fuel to supply the plant. In order to make this model as useful as possible, all future costs are also converted to the equivalent cost today by using a parameter known as the discount rate. The discount rate is almost the same as the interest rate and relates to the way in which the value of one unit of currency falls (most usually, but it could rise) in the future. This allows, for example, the cost of coal purchased 20 years into the future to be converted into an equivalent cost today.

The economic model is called the levelized cost of electricity (LCOE) model. It contains a lot of assumptions and flaws but it is the most commonly used method available for estimating the cost of electricity from a new power plant.

When considering the cost of new power plants the levelized cost is one factor to consider. Another is the overall capital cost of building the power station. This has a significant effect on the cost of electricity but it is also important because it shows the financial investment that will have to be made before the power plant generates any electricity. The comparative size of the investment needed to build different types of power stations may determine the actual type of plant built, even before the cost of electricity is taken into account. Capital cost is usually expressed in terms of the cost per kilowatt of generating capacity to allow comparisons between technologies to me made.

When comparing different types of power station there are other factors that need to be considered too. The type of fuel is one. A coal-fired power station costs much more to build than a gas-fired power station but the fuel it burns is relatively cheap. Its price rarely changes dramatically either. Natural gas is more expensive than coal and it has historically shown much greater price volatility than coal. This means that while the gas-fired station may require lower initial investment, it might prove more expensive to operate in the future if gas prices rise dramatically.

Renewable power plants can also be relatively expensive to build. However, they normally have no fuel costs because the energy they exploit is from a river, from the wind, or from the sun and there is no economic cost for taking that energy. That means that once the renewable power plant has been paid for, the electricity it produces will have a very low cost. All these factors may need to be balanced when making a decision to build a new power station.

CAPITAL COST

A coal-fired power plant is a complex processing plant and it requires a range of expensive materials in its construction. Special steels will be needed for high-temperature components such as the boiler and the steam turbine. Corrosion-resistant steels will be needed elsewhere to resist the effect of high-temperature steam that is often laden with acidic chemicals. Moreover, while some components such as the steam turbines and the generator can be built in a factory and then shipped to the site, much of the construction has to take place on-site. This construction will necessitate a large work-force so labor costs will become an important element of the overall cost.

Labor costs vary widely from country to country and this can make the cost of construction of the same plant quite different depending upon where it is built. The cost will also be affected by commodity prices because of the amount of iron, copper, and other commodities needed in its construction. Today, these are determined by global market forces. This makes the cost of a coal plant sensitive to economic cycles of activity.

Table 8.1 shows figures for the capital cost of a pulverized coal-fired power plant in the USA from the beginning of the twenty-first century until 2014. The figure in the table is called the overnight cost because it does not include any element related to the cost financing any loans needed during the construction of the plant. The capital cost in 2001 was $1011/kW. By 2014 that had risen to $2734/kW. This inflation of the cost of building such a plant is strongly related to commodity prices and the steepest rise in the cost actually took place between 2007 and 2011, a period that saw very high global commodity prices.

The prices in Table 8.1 reflect those in a mature developed market. Costs are lower in other parts of the world. In China, for example labor costs are likely to be lower and the cost of local steel may be lower too.

The figures in the table are for plants without any carbon capture measures. When carbon capture is added to the cost of a coal-fired power station, the capital cost is likely to double compared to the cost without carbon capture. Estimates vary for the different types of

Table 8.1 Annual Capital Cost of a Pulverized Coal Power Plant in the USA	
Year	Overnight Capital Cost of a Pulverized Coal Power Plant ($/kW)
2001	1021
2003	1079
2005	1134
2007	1206
2009	1923
2011	2625
2013	2694
2014	2734
Source: *US Energy Information Administration Annual Energy Outlooks 2001–2014.*	

carbon capture technology discussed earlier but recent figures suggest that post-combustion capture may be the most cost-effective method of adding carbon capture to a coal-fired power station.

FUEL COSTS

For a coal-fired power station the cost of the fuel is probably the most important factor affecting its economics. Coal has traditionally been considered a cheap source of electricity and its ready availability has made it popular in many parts of the world. Most coal is consumed locally and so unlike oil or natural gas the price is normally set locally. In addition, the cost of coal has traditionally be quite stable, although the recent global economic cycle saw an unusually dramatic rise in the cost.

As an illustration of coal costs, Figure 8.1 show the cost of a ton of US Central Appalachian coal, a coal suitable for power generation, between 1990 and 2013. In 1990 one ton cost $31.6 and although there were fluctuations, the cost remained relatively stable throughout the succeeding decade. However, from 2000 costs started to rise significantly, reaching a peak of $118.8/ton in 2008 before falling back to around $70/ton in the middle of the second decade of the twenty-first century.

This unusually sharp peak in coal prices was a global phenomenon with prices in Europe and Asia exhibiting a similar trend. However the

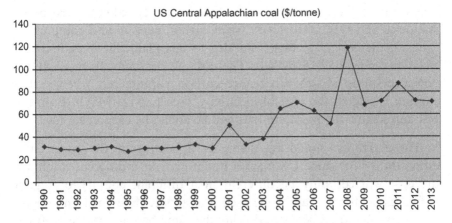

Figure 8.1 The average cost of power plant coal in the USA. Source: BP Statistical Review of World Energy 2014.

Table 8.2 The Levelized Cost of Electricity (LCOE) from Coal-Fired Power Plants in the USA	
Type of Power Plant	Levelized Cost of Electricity ($/MWh)
Pulverized coal-fired power plant	66
Pulverized coal-fired power plant with carbon capture and compression	151
IGCC plant	102
IGCC plant with carbon capture and compression	171
Source: *Lazard's levelized cost of energy analysis – Version 8.0, Lazard 2014.*	

prices of a ton of coal in different parts of the world still vary significantly with local conditions having an important effect on the overall cost. In general, countries that import coal for power generation pay the highest price, a reflection of the high cost of transporting the fuel.

THE LCOE FROM A COAL-FIRED POWER STATION

The capital cost and the fuel cost are the two most important elements in a LCOE calculation for a coal-fired power station.[1] The LCOE in the USA for two types of coal-fired power station, a pulverized coal-fired plant and an IGCC plant, are shown in Table 8.2. Based on

[1] Financial costs can also be significant, depending upon the rate at which interest is charged on any loan taken to finance the project.

these figures, the cost of electricity from a pulverized coal-fired power station is $66/MWh. However, when carbon capture and compression (but not the cost of storage) is added to this plant, the overall cost of electricity rises to $151/MWh. The cost of electricity from an IGCC power plant is $102/MWh before carbon capture but when carbon capture is added, the cost of electricity rises to $171/MWh. Both sets of figures show the premium that will be added by carbon capture.

These figures should be taken as indicative of the cost. Other estimates for the same US market might provide a different result and the cost will certainly be different in a different country or region. If figures of this sort are to be used to compare different technologies, then they should all be derived using the same base set of assumptions.

What calculations of this sort do reveal is that adding carbon capture to a coal-fired power station has a major effect on the cost of electricity from the plant. This is likely to push the cost of electricity from a coal-fired power station above that of electricity from some of the major renewable competitors such as wind power and solar power. However the availability of coal and the scale at which coal-fired power plants can be built will still make then attractive where large-capacity additions are required. Over time the major renewable technologies will make inroads into the power market, providing more and more power as they become more competitive and as the drive to eliminate carbon dioxide emissions becomes more powerful. However, the elimination of coal as a fuel to generate electricity is unlikely to be completed this century.

Printed in the United States
By Bookmasters